BIRDS

BIRDS

A VISUAL GUIDE

Joanna Burger

FIREFLY BOOKS

A Firefly Book

Published by Firefly Books Ltd. 2006

First printing

Publisher Cataloging-in-Publication Data (U.S.)

Burger, Joanna

Birds : a visual guide / Joanna Burger.
[304] p. : col. photos., ill. ; cm.
Includes index.

Summary: An illustrated reference to bird species from all over the world, including information on biology, behavior, distribution, habitat, adaptations for survival and human impact on bird populations.

ISBN-13: 978-1-55407-177-7
ISBN-10: 1-55407-177-1

1. Birds — Pictorial works. 2. Birds — Identification. I. Title.

598 dc22 QL674.B87 2006

Library and Archives Canada Cataloguing in Publication

Burger, Joanna

Birds : a visual guide / Joanna Burger.
Includes index.

ISBN-13: 978-1-55407-177-7
ISBN-10: 1-55407-177-1

1. Birds. I. Title.

QL673.B86 2006 598 C2006-901931-2

Published in the United States by
Firefly Books (U.S.) Inc.
P.O. Box 1338, Ellicott Station
Buffalo, New York 14205

Published in Canada by
Firefly Books Ltd.
66 Leek Crescent
Richmond Hill, Ontario L4B 1H1

Conceived and produced by Weldon Owen Pty Ltd
61 Victoria Street, McMahons Point
Sydney, NSW 2060, Australia

Design: Mark Thacker/Big Cat Design
Design Assistant: Alex Stafford
Cover Design: Jacqueline Hope Raynor
Cover Layout: Sideways Design
Picture Researcher: Joanna Collard
Illustrators: Tom Connel and Mick Posen/The Art Agency
Information Graphics: Andrew Davies/Creative Communication
Color reproduction by Chroma Graphics (Overseas) Pte Ltd

Printed by SNP Leefung Printers Ltd
Printed in China

A distinction between individual species and groups of birds has been made in this book. When an individual bird species is named, such as the Emperor Penguin, the common name is capitalized. When a general group of birds is named, such as penguins, the group is not capitalized.

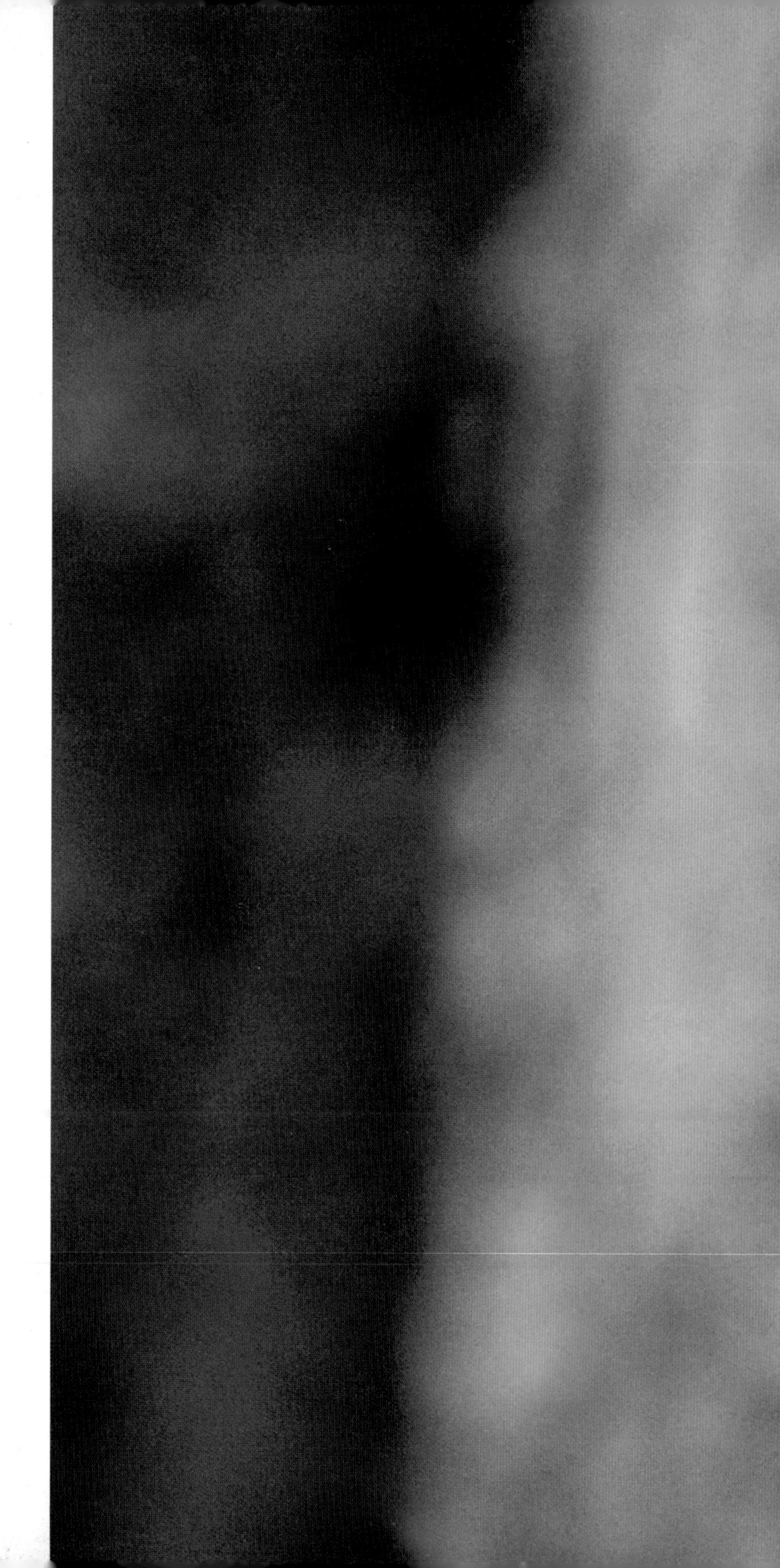

Contents

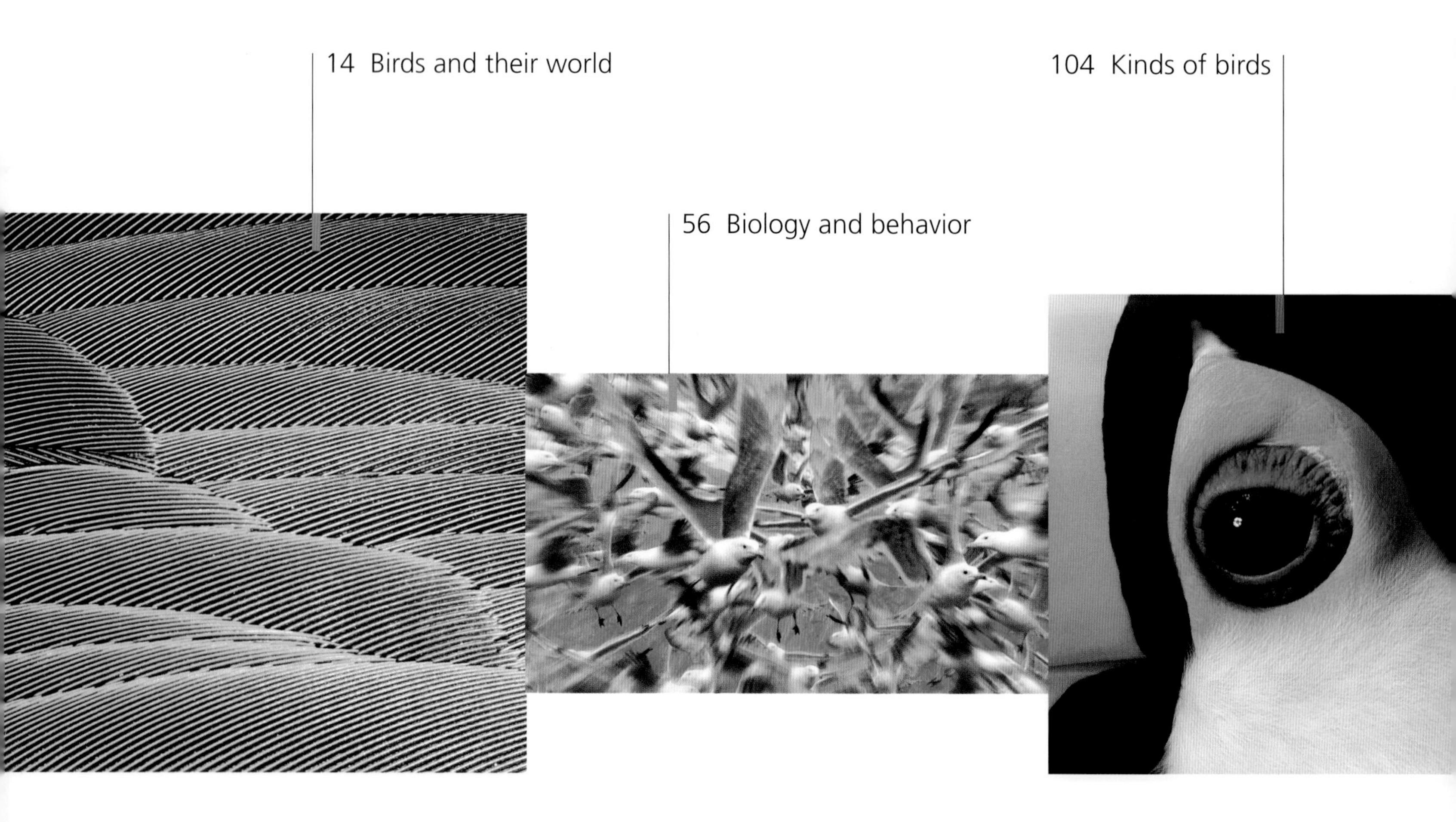

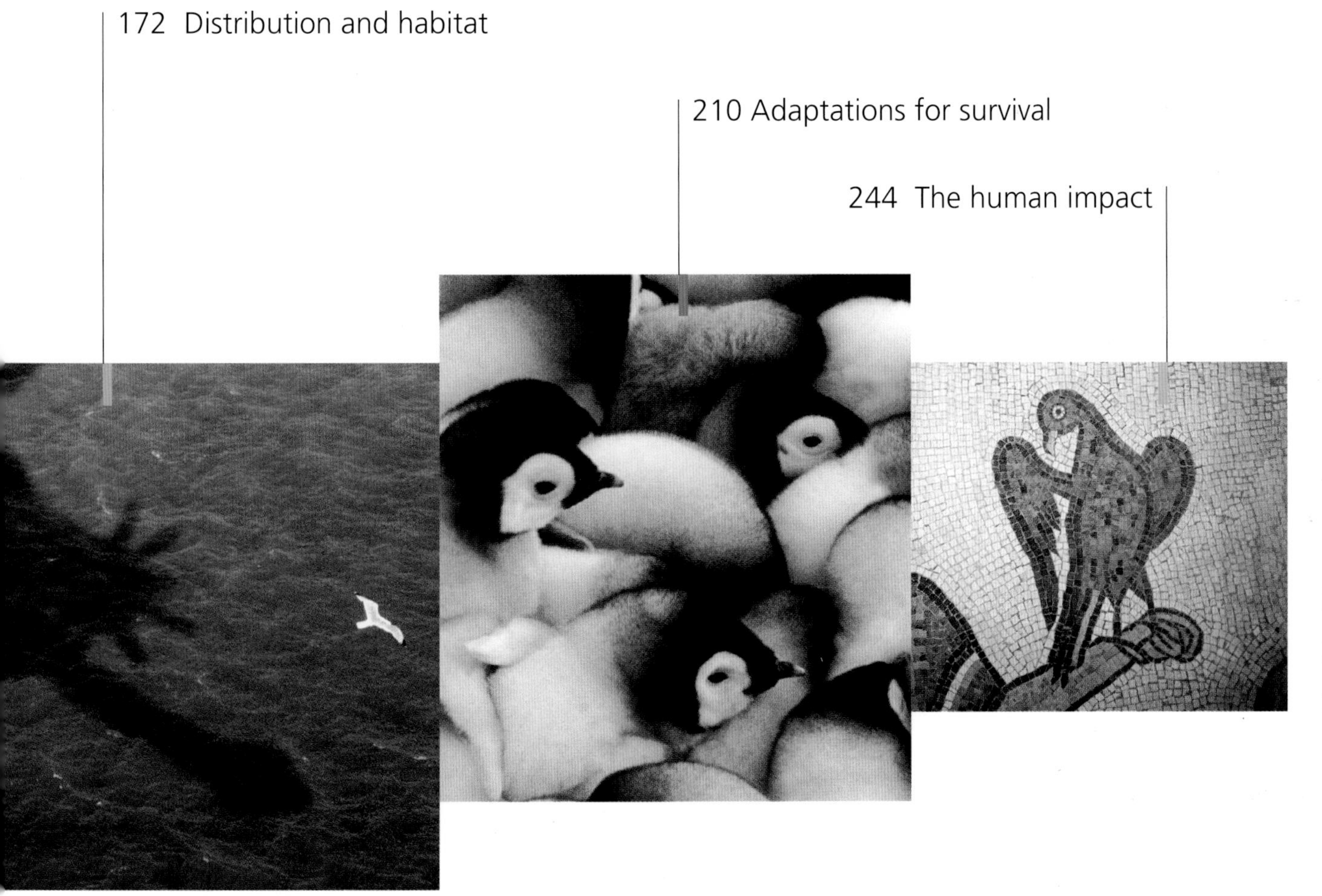

Introduction

The air belongs to birds, although they are also masters of the land and sea. No other group of animals ventures so high in the air, nests on all the landmasses, and dives to such great depths below the water.

Birds have fascinated us as far back in our history as we can explore. Early humans painted them on cave walls and etched them on rocks; the Egyptians created inlaid images of ibises, geese and eagles on the walls of their pharaohs' tombs; native people nearly everywhere in the world emulated their courtship dances and used their feathers in ceremonial dress.

Birds have changed over time, and some have gone extinct, challenging our ability to classify them and understand their evolutionary relationships. Today, many birds are under threat or endangered. Conservation is a global issue that requires international cooperation. Practical measures, however, are controlled predominantly at local and regional levels and they depend on public support and commitment for their success.

Anyone can study the wonder of birds, from backyard bird-watchers to university scientists. The more we learn, the more fascinating birds become. So join us on a journey through the world of birds, from their beginnings as small competitors surrounded by giant dinosaurs, to their divergence to nearly 10,000 different species living in a wide range of habitats, exhibiting diverse plumages and lifestyles. Delight in these masters of the sky and explore their amazing adaptations, fascinating courtship and mating behavior, and incredible beauty.

Birds and their world

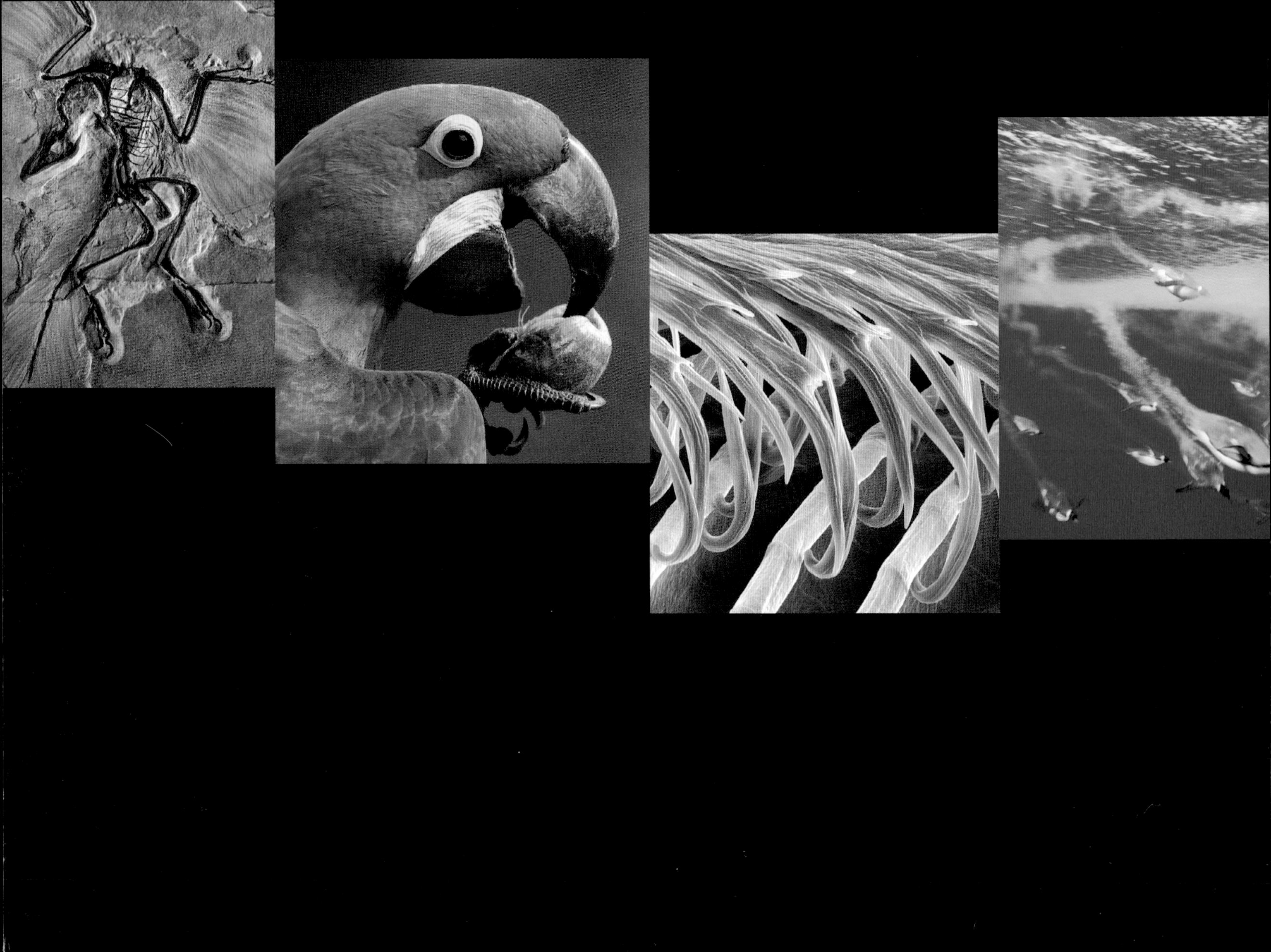

Birds and their world

Descended from reptiles that developed the ability to fly, birds are among the most mobile of animals. In size they range from the tiny Bee Hummingbird to the imposing Ostrich. Among their nearly 10,000 species are birds of almost every color, with a dazzling array of patterns.

What is a bird?

Birds are warm-blooded, two-legged vertebrates and are the only living animals with feathers. These feathers, which provide insulation from heat, cold and water, can be brightly colored to attract mates, or dull to protect from predators. The strong, broad and lightweight wing and tail feathers enable most bird species to fly. Scaly legs and feet are reminders of birds' reptile ancestors. The almost 10,000 species of birds display a huge diversity of colors, shapes and sizes. Ostriches stand taller than humans, while hummingbirds are barely larger than coins. Birds exist almost everywhere on Earth. They have adapted to nest in a wide range of habitats, from harsh dry deserts to wet tropical rain forests, and from Antarctica to the high Arctic. All birds have bills, and all birds lay eggs. In most cases, one or both parents incubate the eggs, although a few species bury their eggs under decaying vegetation to keep them warm.

FLYING MACHINES

Flight distinguishes birds from most other animals. While some other species, such as bats and some insects, can also fly, birds have adapted to flight in a unique way. These feathered flying machines use wings powered by strong flight muscles to take to the air and propel themselves through it. But the rest of the avian body is also adapted to fly: the hollow, lightweight bones; the numerous air sacs that permit increased oxygen flow to the blood; and an enlarged sternum, or breastbone, that allows extra space for anchoring the large flight muscles. Some bones in the skeleton are fused to provide strength for powerful or sustained flight. Backward projections on the ribs overlap other ribs to strengthen the walls of the chest and create extra support during flight. Many birds migrate thousands of miles each year; a few fly more than 2500 miles (4025 km) non-stop across oceans or deserts. Although flying is the chief mode of travel for most birds, some species are flightless.

→ **This dove body feather** shows barbs, running diagonally from left to right, each with an array of barbules on either side. On one side, the barbules end in hooks, shown in purple. These secure the barbules that protrude from the next barb.

← **A bird flies by moving its wings** up and down. This constant movement both projects the bird forward and keeps it airborne. In this series of slow-motion pictures of a flying Java Dove, it is clear that the downward stroke, which takes more energy, gives lift to the body.

← **Hair provides insulation** in mammals. Here, human hair shafts grow from the skin, each one anchored in its own hair follicle below the surface. Hair consists of a fibrous protein, keratin, the same protein that comprises feathers.

←← **Scales protect sharks' bodies**, just as feathers protect birds'. The bony lower part of each scale anchors into the skin. By disrupting turbulence, scales allow sharks to swim without drag.

The evolution of birds

The origin of birds has long been a much-debated issue, although most researchers have agreed for some time that birds evolved from ancient reptiles. But were these ancestors dinosaurs? And were the first birds dinosaurs? Until fairly recently, the fossil record from the Mesozoic era was quite sparse. Even the discovery of *Archaeopteryx*, the earliest known fossil bird, did not resolve the problem. However, recent discoveries of some spectacular Jurassic and early Cretaceous fossils in China provide strong evidence that birds evolved from a group of dinosaurs known as maniraptoran theropods. Birds and maniraptorans share a number of common skeletal traits. As well, studies of these maniraptor fossils suggest that brooding behavior, feathers, flight and the capacity to lay eggs—combined traits once thought to belong uniquely to birds—also occurred in maniraptoran dinosaurs. This supports the classification of birds as theropods, as does the fact that theropods and birds existed at the same time. Before the great extinction of dinosaurs at the end of the Cretaceous period, several different avian groups, including chickens, ducks and ratites, already existed. The study of bird evolution after the Cretaceous has centered largely on the use of biochemical techniques to compare skeletal and other morphological traits.

Birds evolved from a birdlike reptile, such as *Sinosauropteryx prima*, to a wide range of reptile-like birds. Some of these early birds walked on the ground and were quite stocky, but others flew.

← **Modern birds have evolved** widely varied lifestyles and forms of social organization. Parrots, such as this Kea from the New Zealand mountains, are highly intelligent and have long life spans.

→ ***Archaeopteryx lithographica***, from the Jurassic, is the earliest known fossil bird. The original fossil *Archaeopteryx*, discovered in 1861 in a German quarry, is in London's British Museum.

↓ **Hoatzins live along Amazonian and Orinoco tributaries** in South America. Among many primitive characteristics, they have claws on the bend of their wings, which they use for climbing in trees.

EARLY BIRDS

Recent fossil discoveries from the Yixian formation in Liaoning, in northeast China, have dramatically changed our understanding of bird evolution. The Yixian formation has been dated from 145 to 125 million years ago. As well as adult skeletons, researchers there found bird embryos apparently still scrunched in an egg. These fossils provided missing links between theropod dinosaurs, in this case *Sinosauropteryx prima*, an ancient birdlike reptile, and modern birds. In between was a range of early birds, including *Confuciusornis*, *Sapeornis* and *Archaeopteryx*. *Archaeopteryx* has not been found in China. *Confuciusornis* was the size of a pigeon and had long tail feathers. It evolved from feathered dinosaurs. *Sapeornis*, the largest known Mesozoic bird, had claws on the bend of its wing. *Archaeopteryx* was smaller than a small turkey.

Birds throughout time

Archaeopteryx is considered to be a possible ancestor of birds. In response to a range of ecological conditions, descendants of *Archaeopteryx* have diversified in form and function. Adaptive radiation, the process of evolving different species, resulted in great variations in bird features. Diet determined the nature of most bills; diet and habitat influenced the evolution of legs and claws; foraging and flying needs dictated wing forms. All these factors contributed to body shapes. Birds began evolving from bipedal reptiles more than 150 million years ago. Recognizable birds had evolved by the end of the Jurassic period, 145 million years ago, but it was not until the beginning of the Paleocene era (65 million years ago) that modern birds emerged. Released from competition when dinosaurs became extinct, they occupied some of the niches of the bipedal dinosaurs. Modern birds diverged from one another about 60 million years ago. Loons, auks, gulls, ducks, cranes and petrels invaded aquatic habitats between 54 and 37 million years ago, and when, about 25 million years ago, flowering plants emerged, insect- and fruit-eating species evolved. By between 10 and 5 million years ago, birds occupied most of their present-day habitats. As continents shifted, closely related birds, separated from each other, evolved differently. Climate changes also caused many species to seek new habitats.

→ **Adaptive radiation**, the emergence of different species, occurred slowly in birds. Early in the evolutionary process, birds split into two diverse groups. The first, and much smaller, group comprised ratites, tinamous, waterfowl, guans and pheasants. The second group included all the other avian orders.

EVOLUTIONARY SIMILARITIES

Birds that look alike are often, but not always, closely related. Convergent evolution produces similarities in appearance, behavior and ecology among unrelated birds that live in similar environments—or that have solved problems in similar ways. The meadowlarks of the North American grasslands resemble South American meadowlarks, a closely related group. However, except for the absence of long claws on their back toes, these American meadowlarks also look like the longclaws of the African grasslands, to which they are not closely related. Both groups are ground-dwellers of similar size, and both have streaked backs, yellow underparts and a bold black V-shaped mark on their chests. These shared features reflect the two groups' adaptations to their ground-dwelling lifestyle and the similar, though geographically distant, habitats in which they live. Convergence in body shape as an adaptation to swimming can be seen in Diving-petrels from the southern oceans and Dovekies in the Northern Hemisphere.

→ **Birds that resemble each other**, even closely, may be unrelated. On the other hand, birds that look completely different may well be relatives. Despite their different appearances, the Common Hoopoe (right) and the Blyth's Hornbill (far right) are both members of the order Coraciiformes.

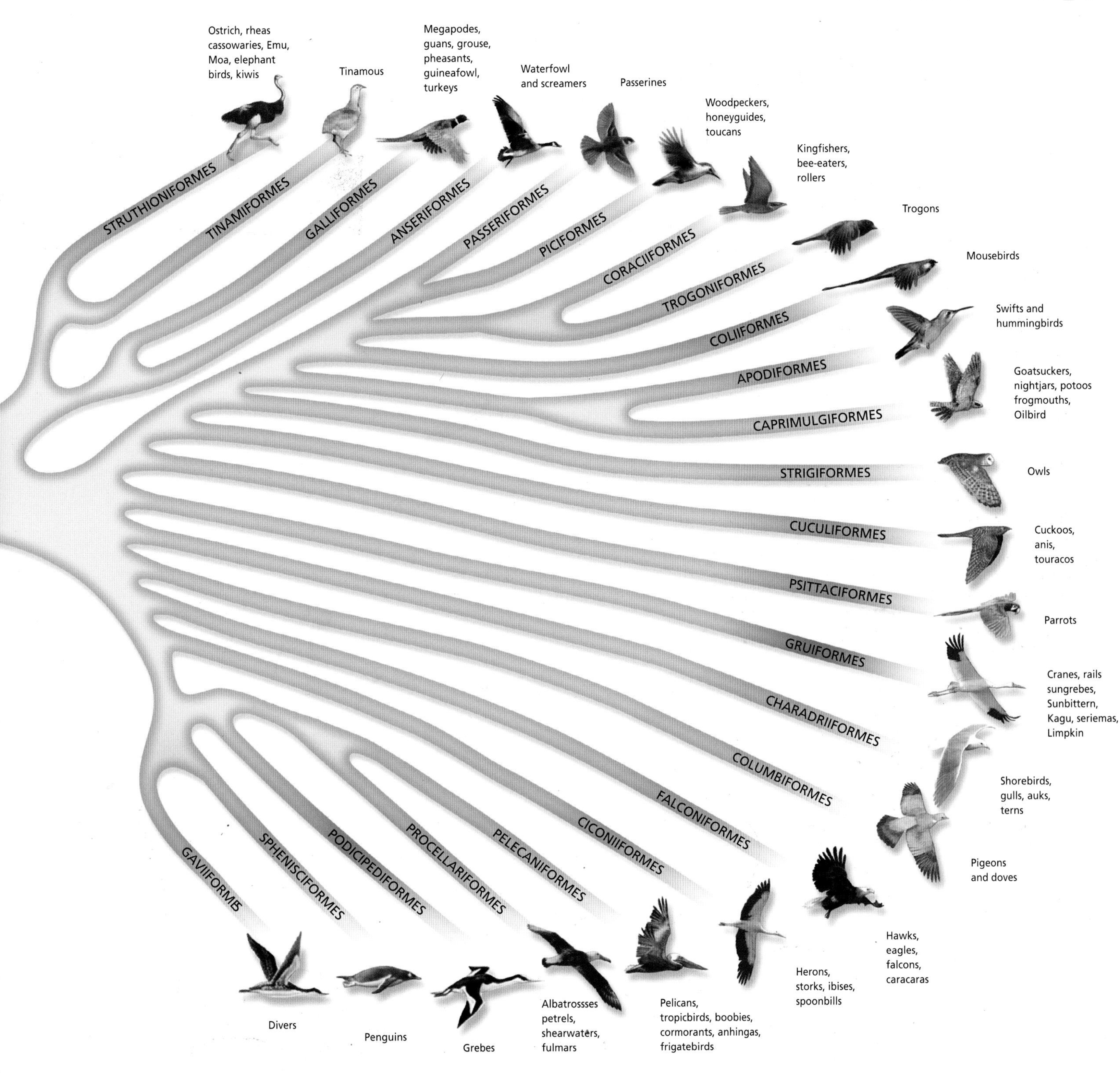
Ostrich, rheas cassowaries, Emu, Moa, elephant birds, kiwis
STRUTHIONIFORMES
Tinamous
TINAMIFORMES
Megapodes, guans, grouse, pheasants, guineafowl, turkeys
GALLIFORMES
Waterfowl and screamers
ANSERIFORMES
Passerines
PASSERIFORMES
Woodpeckers, honeyguides, toucans
PICIFORMES
Kingfishers, bee-eaters, rollers
CORACIIFORMES
Trogons
TROGONIFORMES
Mousebirds
COLIIFORMES
Swifts and hummingbirds
APODIFORMES
Goatsuckers, nightjars, potoos frogmouths, Oilbird
CAPRIMULGIFORMES
Owls
STRIGIFORMES
Cuckoos, anis, touracos
CUCULIFORMES
Parrots
PSITTACIFORMES
Cranes, rails sungrebes, Sunbittern, Kagu, seriemas, Limpkin
GRUIFORMES
Shorebirds, gulls, auks, terns
CHARADRIIFORMES
Pigeons and doves
COLUMBIFORMES
Hawks, eagles, falcons, caracaras
FALCONIFORMES
Herons, storks, ibises, spoonbills
CICONIIFORMES
Pelicans, tropicbirds, boobies, cormorants, anhingas, frigatebirds
PELECANIFORMES
Albatrossses petrels, shearwaters, fulmars
PROCELLARIFORMES
Grebes
PODICIPEDIFORMES
Penguins
SPHENISCIFORMES
Divers
GAVIIFORMES

All shapes and sizes

Birds come in all shapes and sizes, from the tiny Bee Hummingbird that weighs less than one-tenth of an ounce (3 g), to the impressive Ostrich that weighs up to 345 pounds (156 kg). Size and shape vary according to lifestyle. Some birds are adapted for life on the ground, while others are built for long flights. The heaviest birds cannot fly; they simply run through African savanna, walk through forests or skulk through underbrush. The Mute Swan, Wild Turkey and Great Bustard, which weigh up to 30 pounds (13 kg), are the heaviest flying birds. The smallest birds include sunbirds, hummingbirds and kinglets. There are huge differences in the size and shape of birds' bodies, wings, legs, feet, necks and bills. Bills, for example, can be short and thin, short and thick, long and thin, or long and stout. Similar combinations are possible for most body parts. Birds such as grouse and chickens have short legs and round, fat bodies; others, such as herons and egrets, have long legs, long necks and relatively slim bodies. Most songbirds are in-between: they have relatively slim bodies, short necks and tails as long as their bodies. The shape of a bird's wings and tail determines the amount of lift, thrust and maneuverability it will have when flying, just as the shape and size of a bird's bill dictate the kinds of food it can catch and eat.

HABITAT AND FORAGING

A bird's size and shape reflect its habitat use and the way it forages. Because birds are warm-blooded, they must spend much of their waking hours looking for food. Birds that live in tall grasslands—Ostriches, Emus and bustards—often have long legs and long necks. These enable them to forage for seeds and insects while moving easily through the grass and watching for predators. Birds that forage in short grass, such as plovers or sandgrouse, can have shorter legs and shorter necks, and still be vigilant for competitors, mates or predators. Foragers at the water's edge, such as sanderlings, have short legs and necks, and usually run in and out with the waves. Feeders in shallow water, including ibises and spoonbills, have longer legs and necks. Most birds that catch fish in the open ocean have slim, graceful bodies adapted to soaring on ocean updrafts. Albatrosses and petrels can glide for many hours only a few feet above the waves as they search for squid, fish or other prey. Short legs and tails and small, rounded wings permit many other species to forage for insects in shrubs and trees.

Great Bustards are heavy-bodied, flat-headed birds with long legs and necks. Here, a male displays by flinging its wings and feathers away from its body. Male bustards are much larger than females.

A male Ostrich, like the one shown here, can stand 8 feet (2.4 m) tall. With long legs and strong thigh muscles, this bird of the African deserts and savannas can run from danger at speeds up to 43 miles per hour (69 km/h). When threatened, it can deliver a powerful kick to potential aggressors.

← **The Andean Condor**, one of the world's heaviest flying birds, weighs up to 25 pounds (11.4 kg). With a wingspan of nearly 10 feet (3 m), this vulture soars over canyons and high mountains in its search for carrion.

↓ **The tiny Bee Hummingbird** is barely 2.5 inches (6.5 cm) long and is found only in Cuba. Housing and farming expansion has rapidly destroyed much of its nesting habitat, and it is now rare and highly vulnerable.

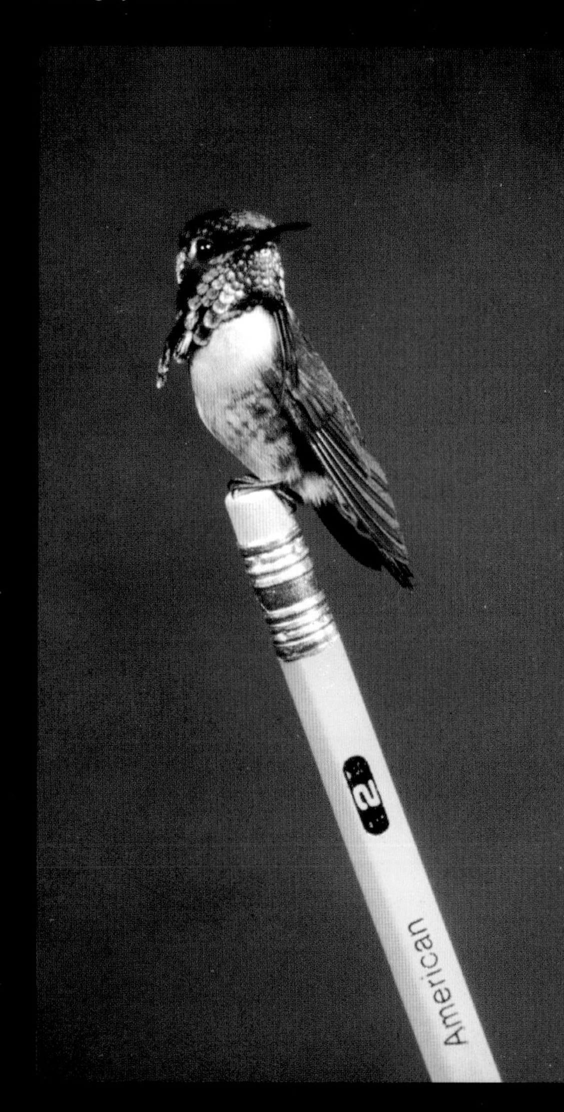

Skeletons and bones

With fewer bones than a reptile or a mammal, a bird is uniquely built to withstand the forces of flight. A bird has special lightweight bones that are fused and reinforced to make its skeleton both incredibly delicate and powerful. The avian skeletal system is made up of bone, cartilage, joints, tendons and ligaments, which, together, support the bird's body. The skeletal system provides the places for muscles to attach, and allows for the storage of phosphorus, calcium and other elements. Bones are living; they slowly change shape and composition in response to such physical stresses as flying or running, as well as to the effects of nutrition, vitamins and minerals. The calcium in one bone today may be in another bone in the future. Some bones also have cavities filled with red marrow, where red blood cells are formed. Bones provide the strength of the skeletal system; cartilage is softer tissue found in joints. As a bird embryo develops, the skeleton is entirely cartilage, which later hardens. This hardening of cartilage into bone—called ossification—occurs in other vertebrates, including humans. Bone can also form directly in tissues without going through a cartilage stage. Ligaments connect one bone to another across a joint, and tendons connect muscles to bones. A bird's skeleton contains two parts: the axial skeleton, made up of the skull, neck, trunk and tail; and the appendicular skeleton, made up of the sternum, the pectoral girdle with the wing bones, and the pelvic girdle with the leg bones. The wing is a modified forelimb.

↓ **Birds are the only animals with a system of air sacs** that increases the surface area of the lungs. Some of these air sacs extend into the bones, making the bones lighter. This increased lung surface area enhances a bird's ability to take up oxygen, but also makes it more vulnerable to airborne toxic chemicals.

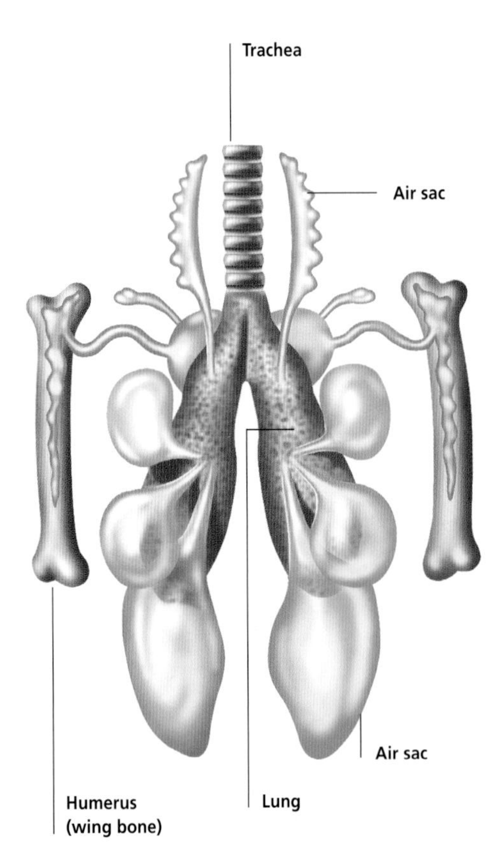

← **Unlike the bones of mammals and reptiles,** the bones of birds are light and airy. This enlarged cross-section of a falcon's skullbone shows the many small internal chambers that make it light without reducing strength or rigidity. Air sacs in skullbones arise from the nasal passageways, rather than the trachea as in other bones. Woodpeckers have less pneumatization in the skull as an adaptation to hammering wood with their bills.

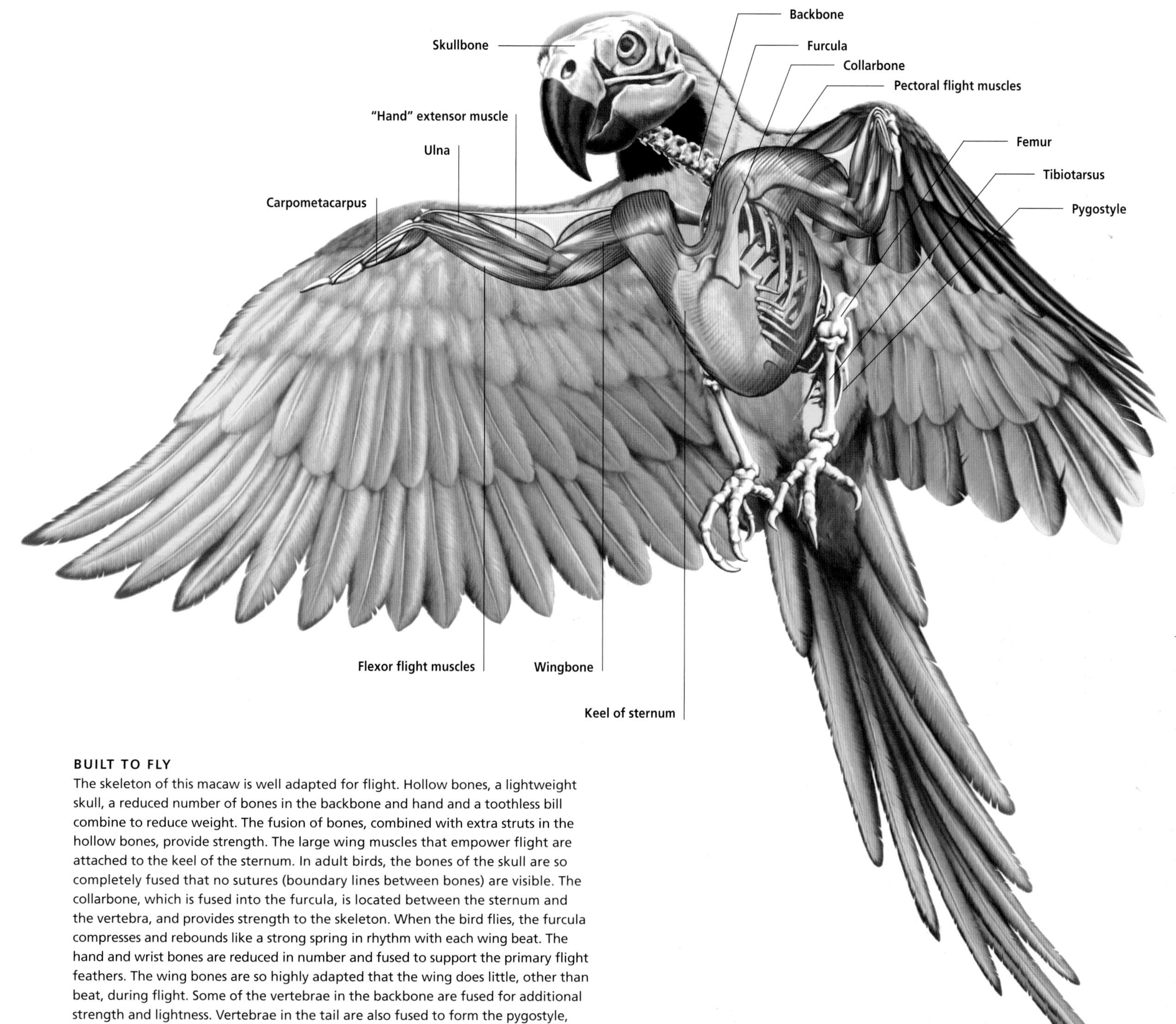

BUILT TO FLY

The skeleton of this macaw is well adapted for flight. Hollow bones, a lightweight skull, a reduced number of bones in the backbone and hand and a toothless bill combine to reduce weight. The fusion of bones, combined with extra struts in the hollow bones, provide strength. The large wing muscles that empower flight are attached to the keel of the sternum. In adult birds, the bones of the skull are so completely fused that no sutures (boundary lines between bones) are visible. The collarbone, which is fused into the furcula, is located between the sternum and the vertebra, and provides strength to the skeleton. When the bird flies, the furcula compresses and rebounds like a strong spring in rhythm with each wing beat. The hand and wrist bones are reduced in number and fused to support the primary flight feathers. The wing bones are so highly adapted that the wing does little, other than beat, during flight. Some of the vertebrae in the backbone are fused for additional strength and lightness. Vertebrae in the tail are also fused to form the pygostyle, which controls the movement of the tail feathers and provides maneuverability.

Inner workings

A bird's internal organs comprise the respiratory, muscular, circulatory, nervous, digestive and endocrine systems. In order to fly, birds need lots of food, oxygen and a means of getting fuel to the muscles. Because flight muscles require so much oxygen, birds' hearts beat much faster than those of mammals. Heartbeat rates range from 70 beats per minute for an Ostrich to 615 beats per minute for hummingbirds. In general, the larger the bird, the lighter its heart will be in proportion to its total weight. Muscles make movement possible and keep all the organ systems working. Birds have three types of muscles: skeletal muscles move the bones, and are what we call the "meat" of a bird; smooth muscles are in blood vessels and organs; and cardiac muscles are the special smooth muscles that make up the bulk of the heart. The nervous system controls all the bird's senses. Nerves transmit sensory stimuli to the brain, which in turn determines all aspects of a bird's behavior.

→ **In South America,** macaws, such as these Red-and-green Macaws, and some other parrots, flock to exposed clay banks to eat clay. Most scientists believe the clay provides nutrients that are missing from their fruit diets.

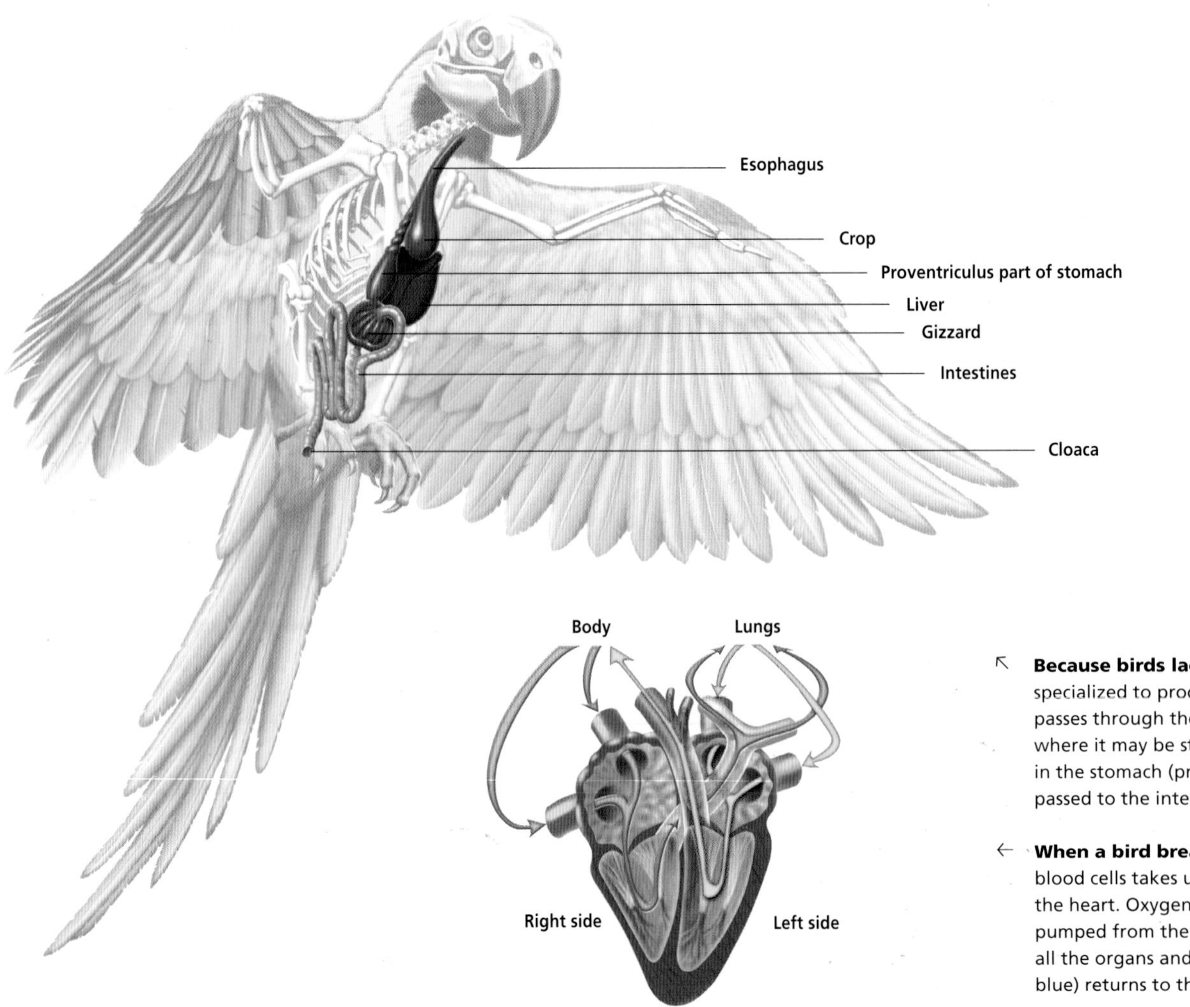

↖ **Because birds lack teeth**, their digestive tract is specialized to process food that is not chewed. Food passes through the bill and down the esophagus, where it may be stored in the crop. It is then ground in the stomach (proventriculus and gizzard), and passed to the intestines where it is absorbed.

← **When a bird breathes**, the hemoglobin in red blood cells takes up the oxygen and carries it to the heart. Oxygen-rich blood (shown in yellow) is pumped from the left ventricle, out the aorta to all the organs and muscles; oxygen-poor blood (in blue) returns to the heart through the right atrium.

PROCESSING FOOD

To prevent weight gain that could hinder a bird's capacity to take swiftly to the air, the digestive system extracts nutrients and energy from small amounts of rapidly processed food. Since modern birds have no teeth, they must either tear food with their bills, eat foods that do not require chewing or rely on digestive juices and the grinding action of the gizzard to break up food. As food enters the proventriculus part of the stomach, gastric glands secrete juices to break down proteins. The food next passes into the thick, muscular gizzard that grinds it to a pulp. In the intestines, food is absorbed through the lining and passed into the bloodstream; waste is excreted through the cloaca. As well as providing nutrients and energy, the digestive tract performs other functions in some species. For example, during the breeding season, doves slough off the lining of the crop and feed it to their young as crop milk. In hawks and owls, the indigestible parts of their prey form into a pellet in the gizzard, and are regurgitated after the flesh is digested.

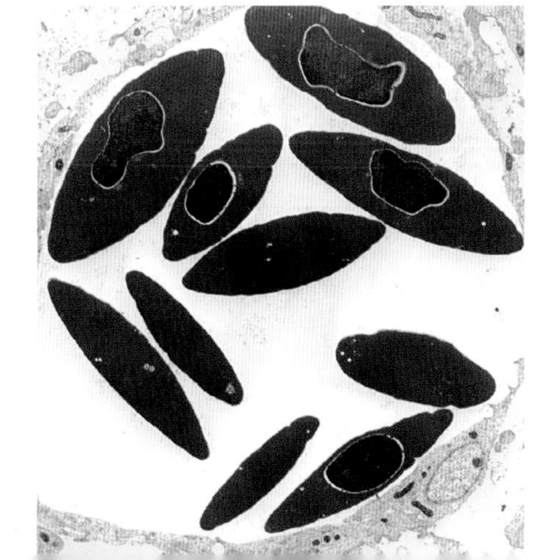

← **Birds' red blood cells** or erythrocytes (shown here greatly magnified) transport oxygen to the heart and around the body. They contain the red pigment called hemoglobin. Unlike a mammal's erythrocytes, which have no nucleus and are round, a bird's erythrocytes contain a nucleus and are elliptical in shape.

Birds' senses

Birds have excellent vision and hearing, although most have a poor sense of smell. Their large and prominent eyes are among the most sophisticated sensory organs of any animals. With such keen sight birds are well equipped to search for food, avoid predators, repel intruders and engage in courtship displays. They can perceive fine detail at great distances, and can see two to three times as far as humans. Birds have an ability to take in a whole scene in an instant, and they have a more highly developed sense of color than humans. Their acute hearing enables them to detect approaching predators or territorial intruders, and to hear the distant calls of mates and potential mates as well as the weak calls of young offspring. The avian ear has three parts: the external ear, the middle ear and the inner ear. The first two sections funnel sound waves from the outside world into the coiled, fluid-filled section of the inner ear (the cochlea). Because of the specialized auricular feathers that protect the ear from air turbulence during flight, it is difficult to see a bird's external ear. Diving birds have stronge protective feathers to shield them from the water. During deep dives, they close the outer ear by folding in the enlarged rear rim of the external ear. Birds can taste and smell, although it is unclear how well they can taste. Many bird species have fewer than 100 taste buds; a human has about 10,000.

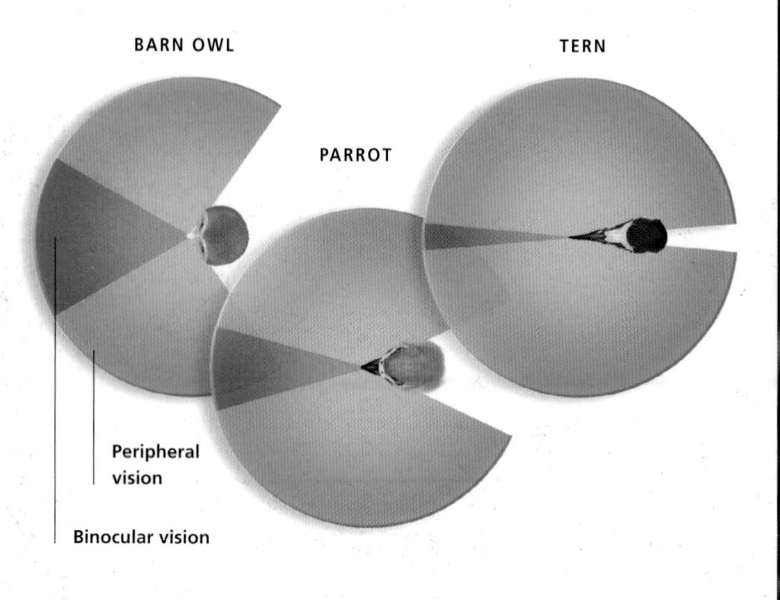

↑ **How a bird sees** depends on the position of its eyes and the shape of its head. Birds such as Barn Owls (right), that have flat faces and forward-looking eyes, have a wide range of binocular vision where they can see an object with both eyes. Birds such as parrots, with eyes toward the front of round heads, have both binocular vision and wide peripheral vision. Other birds, including terns and most passerines, have a fairly narrow field of binocular vision, but wide peripheral vision.

KINGFISHER DIVING STRATEGY

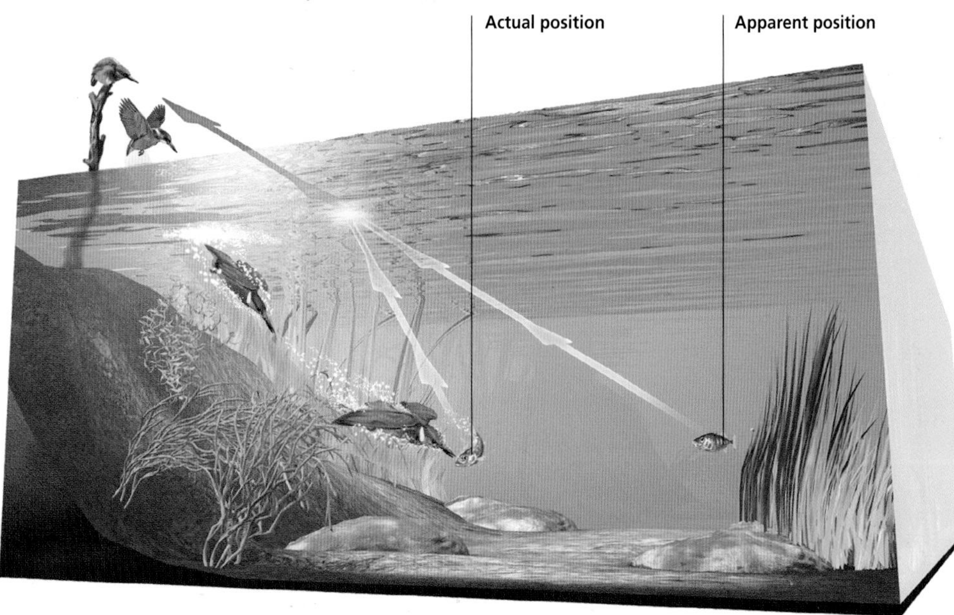

EYES AND EARS

Depending on their lifestyle, different birds have specialized adaptations for sight and hearing. Birds that dive into water for fish below the surface need to adjust their diving so that they are able to catch the prey. If they dove exactly where the fish appeared to them, they would miss. The birds have to account for the water's index of refraction, since an object underwater is actually closer than it appears. There are some interesting specializations for hearing as well. Because the heart-shaped face of a Barn Owl is not perfectly symmetrical, the ears are not at the same level on the horizontal plane. The owl merely tilts its head up and down to pinpoint the location of a mouse moving along the ground, even in total darkness. The left ear is higher than the right ear, and is more sensitive to sounds from below the horizontal. A few birds, including Cave Swiftlets of Southeast Asia, use reflected vocalizations, or echolocation. They find their way through dark caves by emitting short clicks, each click lasting only one millisecond.

↓ **Nestlings, such as these young Great Tits**, have a keen sense of touch well before their vision is fully developed. When they feel one of their parents land on the nest, they immediately open their beaks to stimulate feeding.

↓ **Some birds, such as Turkey Vultures**, have a well-developed sense of smell, which they use to detect rotting carrion. The featherless head is an adaptation that avoids feathers being soiled when the bird plunges its head into carrion.

Bills and tongues

Most birds use their bills for obtaining and manipulating food. Bills vary greatly in size and shape, but in modern birds all are toothless. The size, shape and strength of a bird's bill reflect its diet. Bills are adapted, among other purposes, for tearing meat (hawks), grasping fish (terns), cracking seeds (parrots, finches), probing crevices (woodpeckers), probing in the sand (sandpipers), and straining microscopic food from the mud (flamingos). A bird's bill has four main parts: the upper mandible, or maxilla; the lower mandible; the large jaw muscles; and the horny sheath, called the rhamphotheca, that covers the bill. The rhamphotheca can be smooth, but it has a sharp point in both hawks and parrots, and a serrated edge in mergansers. Birds use their tongues to secure, manipulate, swallow and taste food. Like bills, tongues vary in shape and length. The long, thin tongues of woodpeckers and hummingbirds can delve into tree crevices or tubular flowers, while the forked ends of some hummingbirds' tongues allow them to suck up nectar from flowers. Most birds have salivary glands that moisten the food in the mouth. Bills are usually dull-colored, although some birds have brightly colored bills that they use in courtship displays.

← **Parent European Rollers** push food deep into the bills of their young. This ensures that the young birds swallow the food instead of dropping it onto the floor of their nest. Rollers are related to kingfishers and bee-eaters.

↙ **Like all toucans**, this Toco Toucan from Brazil has a bill almost as big as its body. The bill is very light because the rhamphotheca covers an internal honeycomb of cellular fibers that are light, but strong. Toucans use their bills to pick fruit from trees, prey on small animals, rob other birds' nests and intimidate predators.

↓ **Birds use their bills to bring food to their young**, and also for gathering nesting materials. Here a Jackass Penguin, from Dassen Island in South Africa, brings twigs and roots to its nest. In penguins, males usually do the nest building.

→ **Hyacinth Macaws**, which live mainly in the savannas of Brazil, use their hooked, powerful bills to crack open palm nuts and then drink the milk inside.

SIMILAR BUT DIFFERENT

Even closely related birds can have very different bills. Since the size and shape of a bird's bill determine the foods it can obtain, this variation reduces competition among birds that are otherwise similar. It also helps many species of birds to coexist in the same environment. For example, the shape and length of the bill of hummingbirds that live in tropical rain forests match the shape and length of those birds' main food flowers. A hummingbird with a very short, thin bill can probe only flowers where the nectar is in shallow corollas. Hummingbirds with long, thin, downcurved bills can probe corollas that are deep and curved. The flowers benefit from this specialization, since a visiting hummingbird will most likely move on to a flower of the same species, carrying with it pollen from the previous flower. The birds benefit, too, as there are fewer hummingbirds competing for the same flowers. Similarly, there are many differences between the bills of shorebirds. Plovers have short bills for probing in the sand; turnstones have longer bills for foraging under rocks and deeper in the sand; oystercatchers' bills are longer again and heavy, to delve deep in the mud for shellfish and provide the strength needed to open the shells; and curlews have still longer bills for probing further down in the mud for invertebrates.

Legs, feet and claws

The legs, feet and claws of birds differ almost as much as their bills. Legs and feet can be long or short, and can have or lack prominent claws. In most birds, only the lower leg, known as the tarsus, is unfeathered, although Ostriches' legs are completely unfeathered. The length of a bird's leg reflects both its feeding behavior and habitat. Herons and egrets that feed in shallow water have long legs, while shorebirds that feed at the edge of the waves have short legs. Seabirds that spend little time on land have short legs that are not well adapted to walking, while ground-dwelling birds have long, strong legs for walking, running or dodging through grasses. Although the legs of most birds are black, gray or dark-colored, in some species they are brightly colored. Examples of unusual foot coloring include the brilliant yellow feet of Snowy Egrets. Sometimes the color of their legs and feet intensifies during the breeding season, and then reverts to a dull gray during the winter. Legs change colors, as bills do, when layers of the dead outer skin slough off to expose a replacement layer beneath. In some groups of birds, such as gulls and terns, leg color varies between the different species and is used in identification. The legs of most birds have a hard outer covering, such as scales or leathery skin, that resists injury.

CLAWS AND THEIR USES

The claws at the end of the toes provide clues to a bird's behavior and habitat. Ground-dwelling species, such as larks and pipits, often have elongated hind claws that help prevent them sinking into mud or soft sand. Tree-climbing species, such as creepers and warblers, have curved claws that help them cling onto rough bark and climb on vertical surfaces without slipping or falling. Species that grasp and tear at their prey, such as hawks and owls, have claws that are strong and highly curved. Embryos and young birds of some species have claws on their wings, which disappear as the birds mature. When danger threatens young Hoatzins of South America, they jump into the water and disappear beneath. Once it is safe to return, they use their "wing claws" to climb back up into their nests in the overhanging trees. Unlike other species, Hoatzins retain these claws throughout life, which gives them a rather primitive appearance. Some lapwings also have claws or "spurs" on their wings.

← **Powerful legs, feet and claws** enable many raptors to capture large prey and tear it apart. Ospreys use their feet and long, spike-soled talons to seize large fish from just below the water's surface and carry them to a safe place to eat. An Osprey, which may catch a fish that weighs up to half its own weight, needs considerable strength to carry this prey.

↑ **Some birds use their feet to incubate eggs**. Here, a King Penguin on South Georgia Island holds its single egg on its feet. The feet warm the egg from below and a brood patch provides heat from above.

↖ **A bird's capacity to grasp** will often allow it to perch on small, light objects such as branches, rocks or even blades of grass. Here, a Sharp-tailed Sparrow clings to reeds in a salt marsh of Long Island, New York.

← **Feathers on legs and feet** protect ptarmigans from the cold of the Arctic environment and enable them to walk on deep snow without sinking. In the fall, the Willow Ptarmigan's plumage molts into an all-white coat and is an excellent form of camouflage.

→ **Long, powerful legs** enable many ground-dwelling birds, such as this Emu in Western Australia, to move swiftly. The legs' large surface area also helps the body to shed heat. Emus have three toes per foot, rather than the standard four toes of other birds. They roam open plains in search of fruit, seeds and shoots.

Feathers and plumage

Feathers consist of keratin and usually cover all of a bird's body except the bill, legs and feet. Feathers protect a bird from the heat and cold, provide color for display or camouflage for protection and aid in flight. They grow out of follicles in the skin, in feather tracts called pterylae. Between the tracts are areas of bare skin, called apteria. In a few birds, such as penguins, feathers grow all over the body and there is no apteria, which prevents water from penetrating to the skin. Each feather has a stiff central shaft, called a rachis, and two broad vanes that extend from opposite sides of the shaft. These vanes consist of a series of parallel branches called barbs. At right angles to the barbs, and on the same plane, are branchlets called barbules. By hooking together, these hold the vane intact to form a smooth, continuous and surprisingly rigid surface. When a bird preens, it reattaches the barbs and barbules, which fit together rather like tiny strips of Velcro. Birds molt once or twice a year, depending on the species and type of feather. Large birds have more feathers than small birds. A Ruby-throated Hummingbird has fewer than 1000 feathers, while a Tundra Swan has more than 25,000.

CHANGING COATS

Flight exerts great pressure on feathers and wears them out. Every bird goes through a series of plumage changes during its life. The first feather coat of a young bird may be made up of only a few down feathers. During the first few weeks of life, juvenile feathers push this coat out of its follicles, though wisps of down may remain. Wing and tail feathers grow rapidly at this time, replacing some of the juvenile feathers. The subsequent replacement of juvenal plumage by immature or adult plumage varies from species to species. Some birds grow their adult plumage in the first year; others, such as gulls, terns and albatrosses, take several years to reach adult plumage, and color, patterns. Experienced birdwatchers can age some species by their plumage colors. Adult birds typically molt after breeding. Some birds retain their new feathers for a year, while others may molt again before the next breeding season. In males, bright colors for the breeding season often replace their more somber winter plumage.

↓ **This detail shows the intricate patterns** in the colorful tail feathers of the Germain's Peacock Pheasant, from the humid mountains of South Vietnam. Birds often use their distinct plumage patterns and colors in social displays. The pheasant raises its tail feathers in a wide fan when it displays to females.

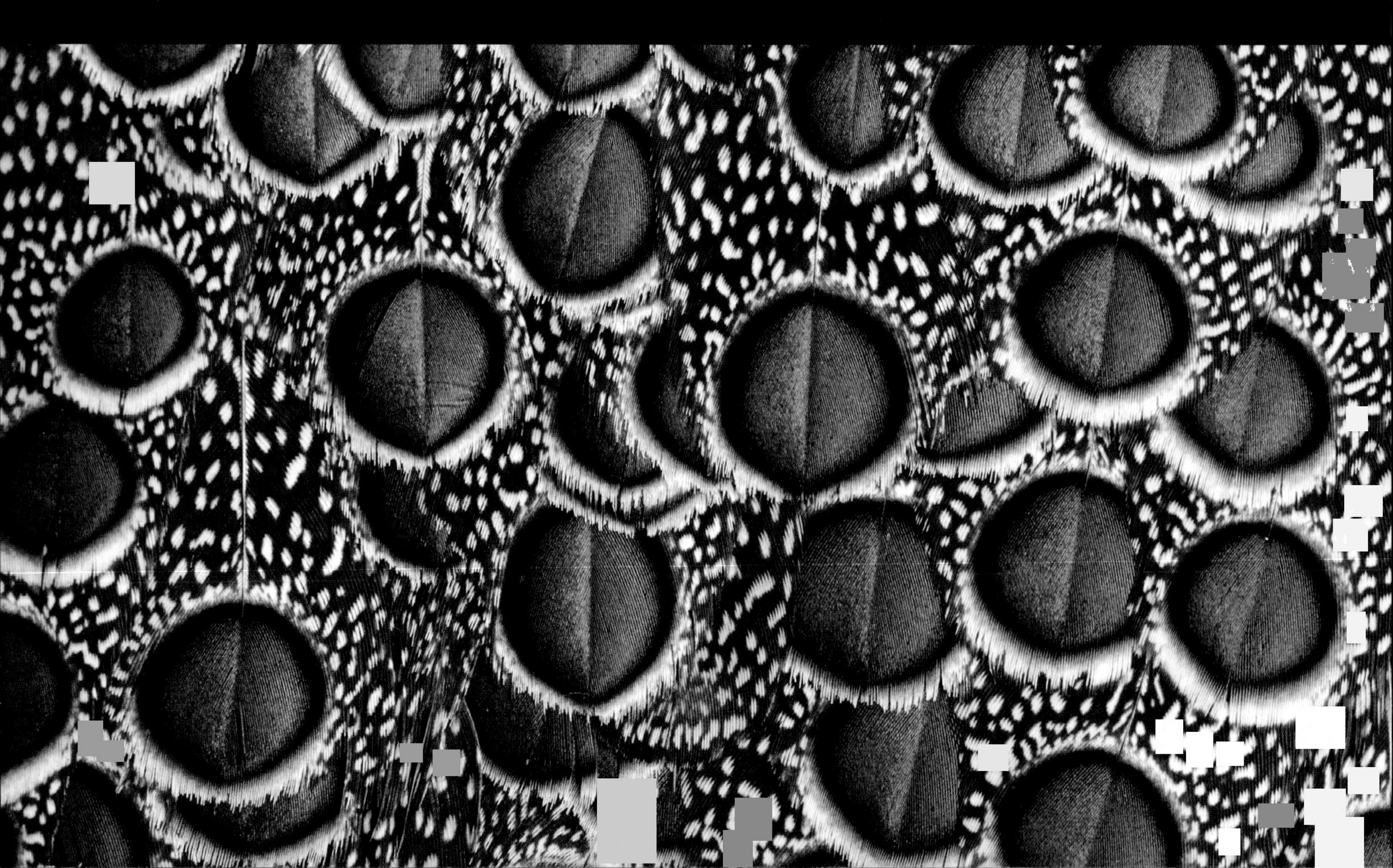

← **The Black-headed Pitohui**, of Papua New Guinea, is the only bird known to have toxic feathers. Its poisonous plumage serves to ward off would-be predators.

↙ **Crested Auklets** are one of several species which have specialized crests, or top-knot feathers, that they use in courtship displays. An insect repellent was also recently discovered in their feathers.

↓ **This enlargement of the vane** of a swallow's feather shows the comblike barbs (round white bands), and the rows of minute, hooked barbules.

→ **A Huli man from New Guinea** wears a traditional headdress surmounted by the tail feathers of a bird-of-paradise. This adult male Raggiana Bird-of-paradise (below right) is fortunate to have survived plume hunters.

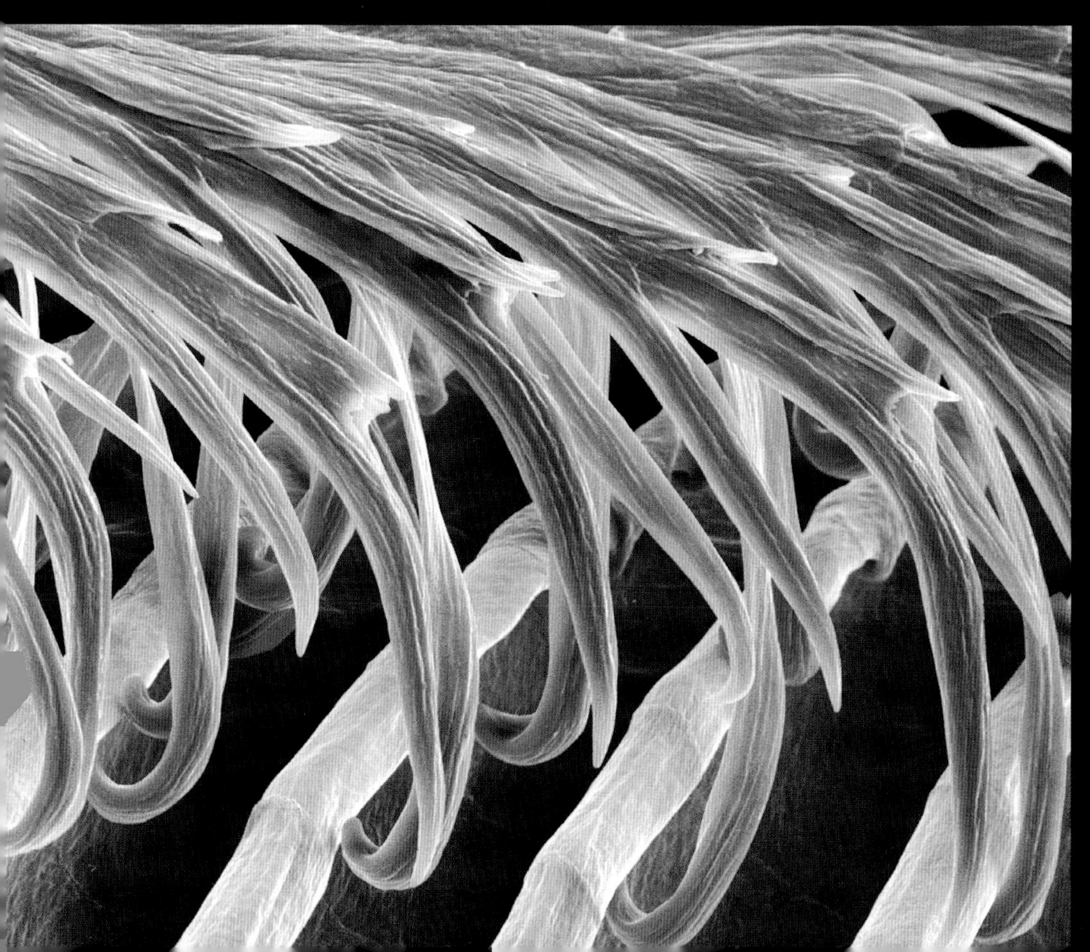

Colors and patterns

Birds' coloration represents a compromise between the brightness needed for courtship display and the dullness needed for camouflage. In birds such as parrots, tanagers and sunbirds, bright reds, greens and blues combine in bold plumage patterns, which suggests that these birds have well-developed color vision. Other birds, however, are drab-colored to blend in with their habitats. Feather colors are produced by carotenoid and melanin pigments deposited in the feather barbs and barbules, and by structural alterations at the feather surface. Most blues and greens are a result of structure; pigments produce the other colors. The colors of birds can vary as a function of sex, age, nutrition and season. In some species, both young and adult birds are the same color, but in others they differ. In some birds, such as gulls, color patterns differ each year until they eventually reach adult plumage at four or five years old. Males and females of the same species can have different-colored plumage. This is called sexual dimorphism. In species where females incubate the eggs alone, they are often drably colored as a form of protection against predators. The reverse applies in phalaropes, where dull-colored males incubate the eggs and colorful females perform courtship. Birds that molt at the end of the breeding season may shed their bright plumage and grow a winter coat of dull feathers that are replaced by brighter ones in the spring. Some feathers that have a role in courtship, such as the plumes of egrets and the long tail feathers of some jaegers, as well as the bright head or facial feathers seen in other species, may be lost at the end of the breeding season.

↑ **In the spring molt**, male Northern Cardinals' red feathers become more brilliant and their beaks become a brighter red. Females remain a duller yellowish red.

← **Wood Ducks exhibit** sexual dimorphism in colors and patterns. Before the onset of the breeding season, the male's dull plumage molts to a bold breeding plumage.

STRUCTURAL COLOR

There are two kinds of structural color: iridescent and non-iridescent. Both result from the microscopic structure of the feathers, which causes only some wavelengths to be reflected. Iridescent colors, such as the throat colors of many hummingbirds and sunbirds, change with the viewing angle. When an iridescent feather is rotated, colors appear and disappear because of the different ways light waves bounce off and go through the melanin granules in the complex barbule layers. The intensity of iridescent color increases with the number of granule layers. Iridescent coloration usually occurs on body, rather than on flight, feathers because iridescent feathers tend to be weaker. Non-iridescent feathers occur when tiny pockets of air, called vacuoles, within cells in the barbs scatter incoming light. Scattering of this sort produces the blue of Bluebirds, Indigo Buntings and Stellar's Jays. Some iridescent colors are the result of reflections from the interface between a feather's melanin granules and keratin layers.

↗ **Little Bee-eaters**, here seen in the Okavango Delta in Botswana, exhibit bright coloration. Males and females are similarly colored. The prominent black lines through the eyes and at the throat actually help to hide the bird in branches. The lines break up the bird's outline and allow it to blend in with its surrounds.

→ **The Resplendent Quetzal**, of the Monteverde cloud forest in Costa Rica, is one of the world's most spectacular birds. Mayans used the long tail plumes of male Resplendent Quetzals, like the one shown here, as ornaments on garments and in ceremonies. They considered it sacred and the Resplendent Quetzal featured prominently in their artwork and legends. Today, it is the national bird of Guatemala.

→→ **The head and casque** of a cassowary are featherless. The skin is brightly colored and the casque is a bony helmet. The red on the neck is part of the long, fleshy wattles. During the breeding season they become inflated and even more brightly colored.

Wings and flight

Birds' wings vary greatly in size and shape. Birds that fly long distances have long, thin wings, while those that maneuver in thick brush have short, rounded wings. The shape of the wing affects the way a bird flies and determines the amount of lift and drag the wing creates when moving through the air. Like the forelimbs of amphibians, reptiles and mammals, birds' wings have three divisions: the upper arm, or brachium; the forearm, or antebrachium; and the hand, or manus. However, the skeleton of the wing is much lighter than that of other vertebrates because the bones are hollow and some contain internal air sacs. The upper arm and forearm of birds are elongated to provide enough space for the attachment of the flight feathers and of the muscles needed for flight. The ratio of a bird's total wing area to its total body weight, known as its wing loading, affects the ease with which it can fly. Some birds, such as Flightless Cormorants, have non-functional (vestigial) wings and are unable to fly.

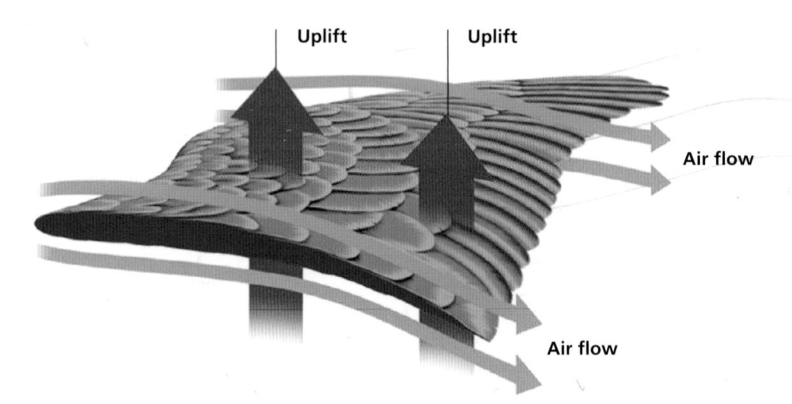

Because birds' wings are more curved above than below, air must flow more quickly across the top as the bird flies. This creates a lower pressure above the wing, known as the Bernoulli Principle, and provides the lift that keeps the bird in the air.

GLIDING AND FLAPPING

Gliding and flapping are the two ways of flying. The soaring or gliding flight of vultures, albatrosses and other birds relies on wing lift to balance the downward pull of gravity. Gliding minimizes the use of powered thrust to overcome the negative effects of drag. Without flapping their wings to thrust forward, birds gradually lose altitude, but they can reduce their rate of descent by using upward air movements. Soaring birds, such as vultures, hawks and eagles, exploit upward columns of warm air, known as thermals. They circle upward within the column of warm air, then glide to the base of the next thermal column and repeat the process. They sometimes glide for miles before entering another thermal column. Other soaring birds, such as albatrosses and petrels, exploit layers of different wind speeds just above the waves. Flapping flight, by contrast, adds thrust to the passive lift. Each flight feather acts as an airfoil. When the airfoils move downward from the horizontal, the generated upward lift propels them forward. Some birds that fly by flapping flap continuously to maintain a steady course; others fold their wings for very short periods to save energy, which produces an undulating flight pattern.

↑ **This swooping Bald Eagle** uses spread wings to slow its descent as it prepares to grab prey in its powerful claws.

→ **A Peregrine Falcon's wings** are adapted for rapid diving. Peregrines are the fastest fliers in the world, and have been recorded diving at more than 200 miles per hour (320 km/h).

← **A Black-browed Albatross glides** on strong air currents over wave-tops. It can soar for hours without beating its wings by using continuous gale winds. This is possible only where winds are strong, as in the "roaring forties."

↞ **Some smaller birds**, such as this Blue Tit, must flap continuously in order to stay in the air and propel themselves forward during flight.

Birds' tails

Birds' tails are generally classified as long or short. A long tail is one that is clearly longer than the rest of the body, while a short tail is the same length or shorter. In some species, such as grebes, the tail is virtually absent. Tails can be square, as in the Sharp-shinned Hawk; rounded, as in the American Crow; graduated, as in cuckoos; pointed, as in the Ring-necked Pheasant; or forked, as in many species of terns. The flight feathers of the tail, known as rectrices, are attached to the fused caudal vertebrae, called the pygostyle. Birds use their tail feathers to steer and brake during flight. Most birds have 12 tail feathers, although some species have more. Anis and grouse have 18 tail feathers and snipe have 24. Birds with especially long tail feathers, such as peacocks, lyrebirds and birds-of-paradise, have differing numbers of tail feathers. Drongos and motmots are among the few bird species that have just one or two very long feathers in the tail. Only males of the species have these long tails, which they use in courtship and territorial displays. However, these conspicuous tails make the males vulnerable to predators and hinder them in flight. In some bird-of-paradise species, only the older males acquire the most elaborate tail plumage. This helps protect more immature males from predators and ensure that they survive to adulthood.

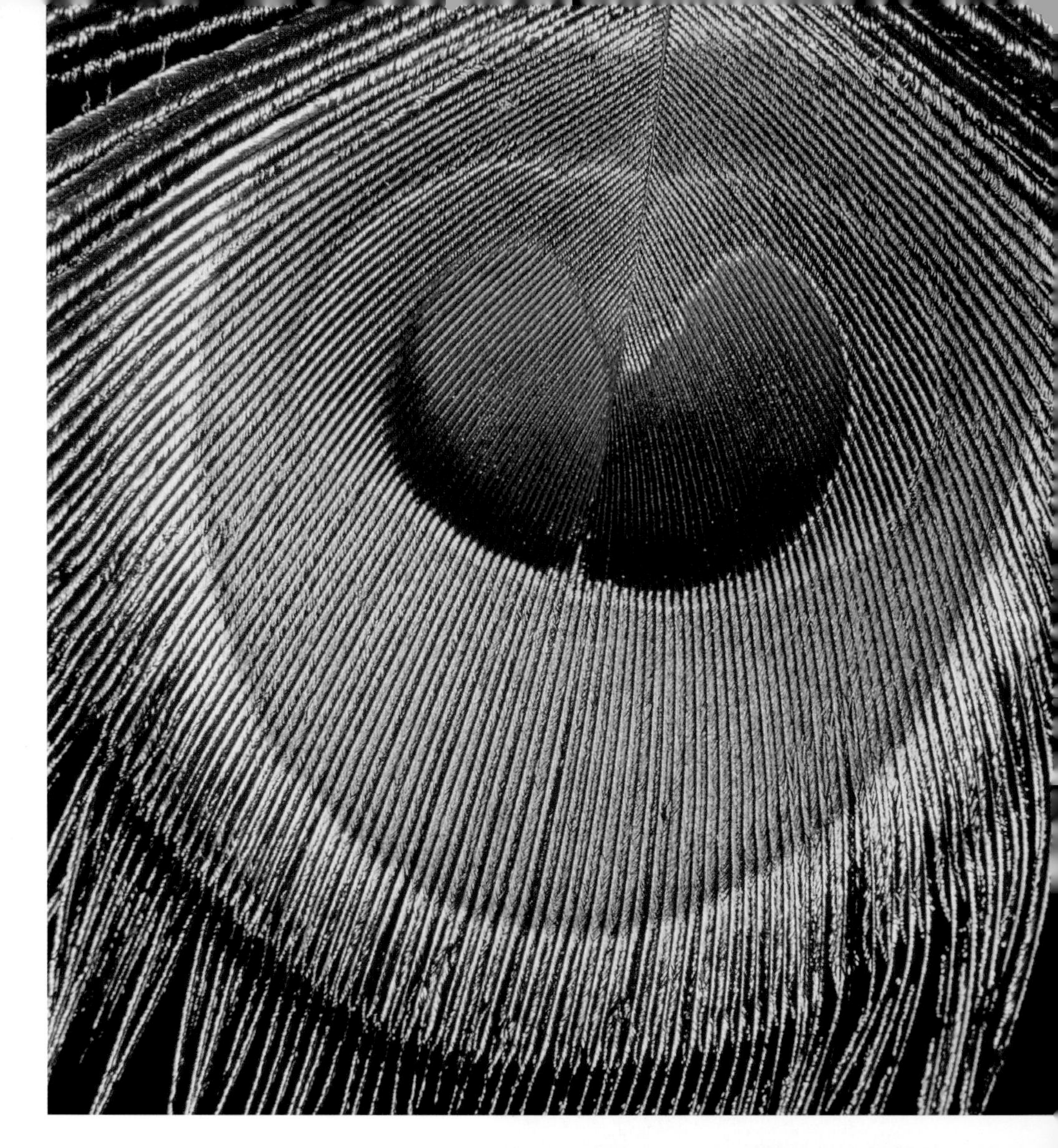

↑ **A male peacock's tail** has more than 200 feathers, each adorned with its characteristic iridescent eye. The bird raises and shakes its tail during courtship and territorial displays.

← **Male Wild Turkeys** have large, iridescent tail feathers that they fan out fully to attract females. As females travel through the woods and forage in small groups, the males follow along, spending most of their day displaying to them.

→ **Wrens have short tails** that tilt upward from their bodies. The Winter Wren (known as the Common Wren in Europe) has the shortest tail of all wrens. It inhabits gardens and boreal forests throughout most of the Northern Hemisphere.

← **Some birds, such as the male Shaft-tailed Whydah**, use their exceedingly long tails in flight displays. Males grow their long tail feathers only during the breeding season. When the display period is over, they shed the long feathers and grow new, shorter ones that are similar to those of the females.

→ **Woodpeckers have relatively rigid tail feathers** that brace against trunks. They help the birds both to climb trees and to balance. Here, a Great Spotted Woodpecker uses its strong, long-clawed toes to grip the bark, and its straightened tail feathers to prop itself against a tree trunk while it probes for insects in the cracks.

← **Motmots in Central and South America** create their own unique tails through preening. The bare shafts and the small fans, seen at the tips of this Blue-Crowned Motmot's twin central tail feathers, are characteristic of motmot species.

TAIL FUNCTIONS

By providing lift and the capacity to maneuver during flight, tails help most birds to fly more effectively. Long tails, which can supply lift, are useful for some species that pursue their prey in flight. Many species, however, have modified tails for other specialized functions, such as for foraging or display. Flightless birds, such as Emus or Ostriches, have abbreviated tails. So too do many species of small birds, such as the Bobwhite Quail and Pygmy Tyrant, that feed and nest mainly on the ground. In these birds, a longer tail would hinder, rather than aid, movement and maneuverability. The elaborate, fanlike tails of some species, such as pheasants and birds-of-paradise, grow naturally into their mature form, but other birds contribute by their behavior to the shaping of their distinctive tails. Motmots have long tails, with two greatly elongated middle feathers that end in rounded "rackets." These feathers grow with vanes along the entire shaft, but the ones closer to the bird's body are weak and brittle, and gradually break off while the bird preens. When perching, motmots twitch their racket-tipped tails sporadically side-to-side.

Crests and crowns

While the wing feathers of most bird species are clearly designed primarily for flight, the feathers on the heads of many birds, like the tail feathers of some species, are specifically adapted for purposes of communication and display. These feathers are often brilliantly colored crests that the birds can erect and sleek back at will, sometimes with great speed. Birds use their crests to attract mates in courtship displays, to deter rivals of the same species or to warn off intruders of other species. The crest feathers can be the same size as other head feathers, or they can be greatly elongated and modified. They encompass a variety of colors, sizes and shapes and occur across a wide range of bird families. In some species only the males have elaborate crests, but in others they are present in both sexes. Other examples of spectacular headgear are the horny or bony protrusions, called casques, on the crowns of some birds, such as cassowaries and some hornbills.

CROWNING DANGERS

Elaborate crown, crest and plume feathers enhance the display variety of birds, but they have also helped to bring some species close to extinction. The Mayans and Aztecs of Mexico and highlanders of New Guinea have traditionally used feathers. Their practice of hunting only a few adult male birds did not greatly endanger any species. Toward the end of the nineteenth century in Europe and the United States, however, it became high fashion for women to wear hats sporting egret plumes. Egrets shed their long, filmy plumes after the breeding season, but plume hunters found it easier to collect the feathers by killing birds in their breeding colonies. Other bird feathers were also targeted and many populations declined markedly. The formation of the Audubon Society—to restore egrets and herons to North America—helped to reverse the trend.

→ **The Gray Crowned Crane** is an African species. Both sexes have strawlike crests that they erect during courtship displays or if they are startled. Courtship rituals involve elaborate bowing and head shaking.

↑ **All New World vultures have bare heads**, which they plunge deep within the bodies of their carrion prey. King Vultures have elaborate coloration on the bare skin of their heads. This plays a role in species identification and during sexual displays.

↑ **Crowned Pigeons** inhabit New Guinea rain forests. They are the world's largest pigeons. Victoria Crowned Pigeons have spectacular crowns of large, fan-shaped lacy feathers, which they erect during courtship or when excited. Their courtship rituals include bowing and display flights. These pigeons suffer great hunting pressure, mainly for their meat.

↑ **The Great Palm Cockatoo** of New Guinea raises and lowers its crest when moved to courtship, aggression, or if it is startled. Its bare face turns a more vibrant red when the bird is excited.

↓ **Common Terns** have white and gray bodies and black crowns. Most other terns also have these distinctive black crown feathers, which in some species can be erected slightly during courtship or during aggressive encounters with other birds.

On the move

The combination of hindlimbs adapted for moving on two legs and forelimbs adapted for flight permits birds to move in many different ways. Most birds can fly, but many are also adept at walking, hopping, running or climbing. Others are adapted for swimming, diving or plunging into water. Birds such as penguins occasionally travel by sliding on their bellies over ice. Birds generally excel at the type of movement required by their foraging and lifestyle habits. American Robins, for example, combine hopping and walking with swift dashes forward as they forage for earthworms. Birds that feed in water have long legs that are adapted for wading. The first flying birds were probably tree-dwellers that hopped from branch to branch when they were not gliding to the ground. Some tree-dwellers learned to climb by hitching themselves up, using their tails for support. When birds first came to the ground, they learned to walk or run by moving their legs alternately. Many terrestrial songbirds hop much of the time. Juveniles of some species hop, and walk only when they reach adulthood. Ostriches and swifts represent the extremes of proficiency in ground movement; Ostriches can run as fast as 43 miles per hour (69 km/h); swifts, the most aerially adapted of birds, can use their weak feet only for clinging to vertical surfaces.

→ **The ability to slow down suddenly** during flight can often be vital for a bird. Here, a Rufous-naped Lark drops and extends its legs and feet, and uses its outstretched, vertically spread wings to slow its flight and land smoothly. Birds with small, short, rounded wings are best suited for short flights.

↓ **Thousands of Lesser Flamingos** taking flight from a lake in Kenya make for a memorable spectacle. These birds literally run across the water until they are airborne. Once in the air, they fly in a large, V-shaped formation. Flamingos fly in long skeins, with their necks straight out in front and their feet trailing behind. Flocks usually travel at night, honking as they fly.

↑ **Emperor Penguins** leave a trail of bubbles behind them as they swim rapidly through Antarctic waters. They often enter the water by tobogganing down icy shores on their stomachs and plunging in. The largest of all penguin species, Emperor Penguins can dive down to almost 2000 feet (600 m) and remain underwater for 20 minutes at a time.

WALKING, SLIDING AND DIVING

Specializations needed for one type of movement may hinder a bird's capacity for other types of locomotion. Birds that move gracefully on or in water are often ungainly and awkward on land. Birds that spend much of their time swimming have lobed or webbed feet that act as paddles to propel them forward, but are less well suited to gripping perches. Penguins' feet are set far back beneath their bodies to aid in smooth, efficient swimming. As a result, they have to walk upright on land to maintain balance. When they need to move swiftly, penguins slide, or toboggan, on their bellies, forcing themselves forward by rapidly beating their stubby wings. Some birds, including gannets, boobies, Brown Pelicans and terns, plunge-dive into water to catch fish, moving faster as they descend. The height of the dive depends upon the bird's size and the depths to which it needs to go. Terns, which usually catch fish only a few inches beneath the surface, normally dive from only a few feet above the water. Northern Gannets, by contrast, can plunge-dive from as high as 100 feet (30 m) to reach depths of up to 12 feet (3.7 m). The many minute air sacs under their skin act as shock absorbers and so allow gannets to withstand the impact of these rapid dives.

Nests

Birds build nests to secure their eggs during incubation and to provide a safe place for developing chicks. The size, shape and materials of the nest depend upon its position, the number and size of the eggs, the eventual size of the chicks and the size of the parent birds. When the eggs hatch, the nest will need to hold the fully developed chicks and both parents without falling to the ground or sinking in the water. Nests take many forms. A typical songbird's nest is cup-shaped, woven from grasses, leaves and fine twigs, and placed on the branches of a tree or shrub; skimmers' and terns' nests are shallow impressions in the sand; swallows' nests, attached to bridges and buildings, are made entirely of mud; the nests of grebes are built of cattails and reeds and float on water; weavers and orioles weave elaborate hanging nests; Social Weavers and Monk Parakeets construct huge communal stick nests. Nests have distinctive characteristics—their features often provide clear evidence of the identity of their owners.

↑ **A mere scrape in the ground**, excavated by vigorously kicking sand, is all that the Black Skimmer has for a nest. It relies on camouflage for protection, first of the spotted eggs, and later of the young chicks.

→ **Black-browed Albatross** build rounded nests of grasses, sedges and seaweed, which they cement together with mud and guano (droppings). These large birds nest in colonies on Antarctic and subantarctic islands.

NEST WEAVERS

The male Black-headed Weaver stitches strips of vegetation together to make a roofed basket. He ties knots, using grasses, to make the nest secure. Then he displays in front of the nest to attract a mate. Male weavers are usually polygamous. Weavers nest in colonies, with the females raising their young alone and working with other females in the colony to look out for predators. The seeds that these birds eat are abundant, so the males have no need to defend a territory and food source.

Egg-laying and eggs

Like their reptilian ancestors, birds lay eggs. An egg is a self-contained chamber in which the energy-rich yolk nourishes the developing chick within. Bird eggs are "cleidoic," which means they have hard shells that shield the contents from the surrounding environment. Egg shapes range from almost spherical to elongated. The greatly varied sizes and shapes of birds' eggs represent a compromise between structural needs, clutch weight in relation to the female's body weight, and egg content. Absolutely round eggs have stronger and more enduring shells and conserve heat better. Energy limitations help to determine the number and the size of the eggs a female can lay, as well as the size of a clutch that a parent can incubate. The number of eggs, either fixed or variable, that any species lays is, however, essentially an inherited trait. Longer-lived birds tend to lay smaller clutches. Songbirds and other small landbirds have clutches of between two and twelve eggs; the exact number depends upon the species, food availability during the egg-laying period, the female's age and condition, and the quality and geographical position of the territory. Most shorebirds lay four eggs; most pelicans and petrels lay only a single egg. Many birds lay fewer eggs in their first breeding season than in subsequent seasons.

→ **Brightly colored eggs** that stand out from the surroundings are usually laid in deep nests or well hidden by foliage. In contrast, ground-nesting birds tend to lay dull colored eggs that blend in well with sand, leaf litter and the ground.

↓ **Some cuckoos lay their eggs** in other birds' nests. The cuckoo eggs (on the right in each clutch) are usually slightly larger than the host's eggs, but similar in color.

↑ **Thick-billed Murres** usually lay a single egg on narrow bare rock ledges, on coasts and cliffs in Arctic and subarctic regions. Unlike most eggs, a Murre's egg is pear-shaped. When pushed, it rolls in a tight circle, which prevents it from falling over the edge.

KEEPING EGGS MOIST

The hard shell of a bird egg protects the developing embryo inside. However, no egg is large enough to hold all the oxygen the embryo requires, or to contain all the waste carbon dioxide and water the growing embryo produces. In effect, the developing embryo "breathes" through the shell. So that it remains porous and does not become brittle, the shell must stay fairly moist. Birds that nest in very dry habitats, such as on hot sand on remote oceanic islands, have developed strategies to prevent their eggs drying out. Sooty Terns (shown below) dip their bellies in the ocean and carry water back to the nest to wet the eggs for evaporative cooling. The parents take turns to incubate the eggs. While it is incubating, a parent flies out to wet its feathers, returning quickly to the nest to prevent injury to the embryo from the hot sun. Other birds that nest near water, including shorebirds, gulls and other terns, also wet their nests or eggs.

↑ **An egg from the largest known bird**, the extinct Elephant Bird, sits beside the egg of an Ostrich, the world's largest living bird, and the smallest bird egg of all—that of a hummingbird. An Elephant Bird's egg could hold more than 2 gallons (8 l) of fluid.

SOOTY TERN EGG WETTING

A Sooty Tern dips down toward the waves to wet its belly.

Returning to the nest, it drips water from its belly onto the egg.

Food and drink

Birds spend a good deal of their waking life looking for food and eating it. In order to survive, move about, migrate and reproduce efficiently, they need to get as much nutrient-rich and energy-providing food as they can as quickly, and with as little risk, as possible. Many birds are herbivores that eat only plant material. Other birds are carnivores that consume only fish or meat. Some birds are omnivores that eat both plants and animals. The wide range of foods that birds eat includes fruits, seeds, grasses and other vegetation, algae, crustaceans, mollusks, insects, worms, grubs and other invertebrates, fish, amphibians, reptiles, mammals and other birds. Some birds of prey feed predominantly on carrion. While some species have very limited, specialized diets, others forage much more widely. Even within a species, factors such as age, sex, location, seasonal variations and even individual preferences, can affect what birds eat. Adults and their young can have different food needs. Some adult songbirds, for example, survive on seeds and fruits, but catch highly nutritious insects to feed to their young. Gulls, which scavenge rubbish dumps for their own sustenance, find fish and shellfish for their offspring. Most birds can adapt their diets to changing environmental situations.

DRINK REQUIREMENTS

Food provides energy and nutrition. Water, too, is essential for maintaining life processes and facilitating the elimination of wastes. Most birds satisfy their fluid needs by drinking, as often as they need to, directly from rivers, creeks and dams. Birds that eat fruit, fish, worms and insects, or that drink nectar, may obtain much of the water they require from these sources. Seed-eating birds have the greatest need for water and often drink several times a day. Water economy is particularly critical in arid environments, especially in the middle of the day when, to avoid heat stress, birds need to lose heat by evaporation. Desert-dwelling species that must frequent waterholes are particularly vulnerable to waiting predators. Sandgrouse arrive at waterholes just as darkness decends to reduce this risk.

↙ **Vultures are specialist carrion-eaters.** Groups, such as these Turkey Vultures, often feed together at a carcass, diving into the body cavity. Although New World and Old World vultures belong to different families, they exhibit similar carrion-feeding behavior.

↓ **Northern Gannets** live in coastal waters and feed exclusively on fish, such as mackerel, herring, garfish and whiting, that swim not far below the surface. They drop vertically onto prey in towering plunge-dives.

→ **A pair of adult American Robins** fight for possession of a worm. These birds feed on a range of worms, grubs and other insects. Chicks in the nest are mainly fed worms. Both parents are responsible for feeding the young.

↓ **Oceanic birds** that rarely come to the land have specialized adaptations for drinking seawater. Shearwaters, petrels and albatrosses, such as this Wandering Albatross, have large nostrils that extend to the top of the bill and are protected by short tubes. When drinking saltwater, they secrete the salt, which would otherwise poison them, through these tubes.

Brains and intelligence

Birds have well-developed brains and can master complex skills in a range of situations. Relative to their total body size, birds' brains are large—up to eleven times as large as those of similar-sized reptiles. Birds' brains analyze signals, integrate them with remembered experiences, channel them through the neural system and transmit them as motor instructions to the body. The degree of intelligence between species is related to the challenges they face and to their longevity. Birds that live long lives, or that need to cope with changing environments and many competitors and predators, must learn complex foraging and reproductive techniques to survive. As in all vertebrates, the forebrain, responsible for behavioral instincts, sensory integration and learned intelligence, includes the olfactory bulbs and the cerebral hemisphere. The midbrain regulates vision, balance, physiological controls and hormones for reproduction. The hindbrain links the spinal cord to the peripheral nervous system.

← **Crows and ravens** are highly intelligent. They quickly work out how to open backpacks and packages in search of food. Here, a Common Raven uses its beak to open the compartment of a snowmobile. Sometimes ravens rip open compartments or strip off pieces of metal seemingly for the fun of it. They are especially fond of shiny objects.

BIRDS USING THEIR BRAINS

Among the many obvious examples of intelligence are the song-learning mastery of songbirds, the capacity of parrots for language and mimicry, the unique foraging behaviors of many birds and the problem-solving abilities of crows and jays. Birds also develop a sense of numbers more quickly than most mammals. Parrots excel at using their feet to manipulate food and other objects. In the wild, African Gray Parrots can mimic the sounds of hundreds of other birds and animals. They can also communicate using human language. Laboratory experiments have shown that an African Gray can achieve vocabulary levels of a four-year-old human, can categorize objects as well as a chimpanzee, and may have better visual–perceptual strategies than an adult human.

→ **Black Egrets of Africa** trick fish by creating a shadow with their wings. The fish are attracted by the shadow, mistaking it for a riverbank or protective vegetation, and swim straight into it.

Some birds have learned unique ways of attracting prey. Green Herons drop leaves or small sticks into moving streams and then run downstream to snatch and eat any small fish that their "lure" has attracted. They then return upstream to deposit another small stick or leaf. Since these herons carefully select sticks or leaves of a size that will swirl as they travel, this behavior is an example of tool use.

GREEN HERONS FISHING

← **Sometimes a Black Egret** makes a complete "tent" with its wings. It bends its body forward, raises its wings over its back, extends the wing tips downward and tucks its head below its wings. This tent creates an even larger area of shadow and reduces reflections on the water's surface. It improves the bird's vision at the same time as shielding it from the blazing African sun while it stands still and waits for unsuspecting fish. In the early morning, or when it is cloudy, Black Egrets do not make tents because they cannot make use of shadows. Instead, they merely walk along, searching for fish.

Biology and behavior

Biology and behavior

All birds have fundamentally similar anatomical and physiological features. They do, however, display a wide range of nesting, reproductive and social behaviors. The shape, size and structure of their bills are highly varied to enable a diversity of feeding strategies.

Attraction and mating

All birds need to find mates. In most cases, the male initiates courtship, though in a small number of species, such as pharalopes, the female takes the sexual initiative. Ways of attracting mates can include the possession of bright feathers, or other prominent features; the use of feathers, throat sacs or feet in elaborate displays and rituals; the performance of courtship songs and dances; and often prolonged aerial chases and displays. In some species, such as most songbirds, courtship is brief and may last only a few hours or days. In others, the process may continue for weeks, months or even years. Species that form long-term pair bonds generally have more elaborate courtship displays than those whose bonds are more temporary. In some cases, many males gather together in staging areas, called leks, where they compete to attract females. Copulation, or the insemination of the female with sperm, can take place at the end of courtship, or during the courtship period. Some species copulate only once; others copulate ever more frequently until, finally, the females lay the eggs.

→ **Great Egrets** have long, lacy back plumes, known as aigrettes, which they use in courtship displays. Single males often stand on their nest, displaying to prospective mates that might fly over or land nearby.

↘ **Blue-footed Boobies** have vibrant blue feet which they display and lift during their elaborate slow-motion courtship dance.

↓ **A graceful male Japanese Crane** dips and bows to a potential mate at the start of what can develop into a frenetic mutual courtship dance.

MUTUAL INVOLVEMENT

Most ducks court on the water, with the males performing displaying vigorously. Females mainly watch; their role is to accept copulation. Most courtship displays, though, are more mutual affairs. Once a male has attracted a female to his vicinity, both sexes begin their rituals. In many songbirds, males use territory defense, as well as display, to attract females. After a pair bond has formed, songbirds begin copulating and laying eggs. In other species, such as herons and egrets, both sexes engage in bouts of scraping, bowing and bill touching that last for days. Boobies, penguins, grebes, albatrosses and cranes have extensive displays, dances and ballets that may go on for weeks. Male albatrosses select territories before they are fully mature, and give elaborate courtship displays, even with no prospective mates nearby. When a female does appear, the two display to one another for days, or even weeks. If the female leaves, the male resumes his lone display. It may take two to three years before he finds a mate, and another year before eggs are laid. With a potential lifespan of 50 or more years, albatrosses can afford to be patient.

→ **A pair of Western Grebes** skim across the water's surface in a synchronized courtship ballet.

↓ **Male Magnificent Frigatebirds** have brilliant red inflated throat sacs that attract females to their nests.

Lasting and short-term pairing

Pair bonds between birds vary from brief sexual encounters to lifelong pairings. The length and nature of these bondings represent different kinds of mating systems. Hummingbirds, manakins and grouse form short, indiscriminate sexual relationships that we term promiscuous. A pair bond involving multiple mates is polygamous. If males have more than one mate, it is polygyny; if a female has several mates, it is polyandry. Phalaropes are polyandrous. Polygynandry occurs when both sexes pair with several mates of the opposite sex. Monogamy is by far the most common mating system for birds. Nearly 90 percent of all species form prolonged and basically exclusive bonds with a single mate. A monogamous bond, though, may last as briefly as one breeding season or as long as a lifetime. And recent studies have shown that even within monogamous pairs there is some level of extra-pair copulations. Most species display only one kind of mating pattern. Some species of shorebirds, however, alternate between several mating systems.

← **Two Whooper Swans** flap their wings during a greeting or courtship display on the tundra. As well as maintaining lifelong pair bonds, which can span up to 35 years, these swans usually migrate together in family groups.

↙ **Wandering Albatrosses** engage in courtship displays on South Georgia Island. Albatrosses are monogamous and maintain their pair bonds throughout their long lives. Each year they renew their bonds with complex head movements and courtship rituals that can last for days.

→ **Attwater's Prairie-chickens** are promiscuous. Males display by inflating their yellow-orange air sacs, stomping and hooting. Females select males for copulation, and then build nests, lay eggs and incubate on their own.

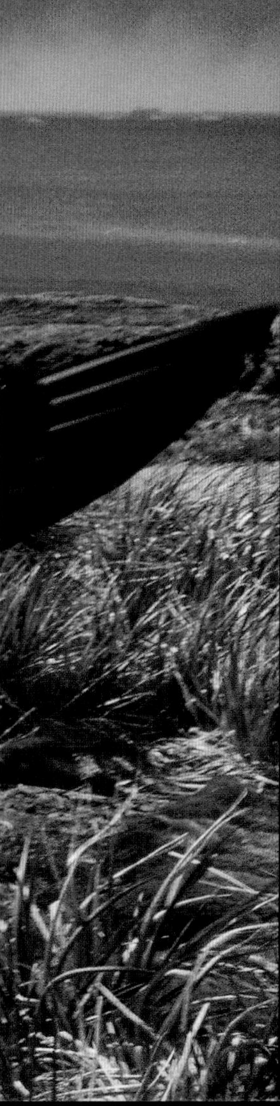

THE CASE FOR MONOGAMY

Most birds are monogamous because that pair-bond system allows them successfully to produce and raise the maximum number of young. In a monogamous bonding, both parents care for the young, sharing responsibility for defense, incubation and feeding of their offspring. In a monogamous pairing, one parent is able to be away from the nest foraging for food while the other remains behind to defend their territory and care for the chicks. Eggs or chicks that are left alone, even for short periods, are vulnerable to thermal stress and to marauding neighbors and other predators. As well, birds that have only one mate do not need to devote time to attraction and courtship rituals, and generally lay eggs earlier than polygamous or other birds that regularly change partners. In general, the earlier birds lay, the greater number of young they will be able to produce and raise. Stable monogamous pairings have obvious advantages over short-term ones for producing and rearing young. Even in long-lived species, mortality can take a heavy toll and force birds to establish new pair bonds. Incompatibility and infertility are other factors that can also affect long-term monogamous pairings.

↗ **Many songbirds, like these Crested Tits,** are monogamous, but they may keep the same mate only for one breeding season. Crested Tits live on average for two years. The short lifespan of small birds such as these means that they frequently need to attract new mates.

→ **Peach-faced Lovebirds** are monogamous and form long-term pair bonds. Like many monogamous birds, they engage in a number of pair-bond behaviors, such as mutual preening, that help cement their relationship.

Nest building

Birds build nests to protect their eggs and newborns from predators and the weather, and to provide warmth for incubating eggs and maintaining the body temperature of their fledglings. Once the young hatch, they usually stay in the nest for some time. Sooty Terns on oceanic islands nest in large colonies, sometimes with up to a million nests, packed so close together that birds in adjacent nests can almost touch beaks with one another. Guanay Cormorants nest in colonies of 4 to 5 million birds, with densities of up to 12,000 nests per acre (30,000/ha). Species that nest on the ground, such as some terns, shorebirds and nightjars, often construct almost no nest, and merely make an impression in the sand or bare ground. Other ground-nesters, however, build elaborate nests. Birds that nest on the water, such as grebes and some gulls and ducks, usually build substantial floating nests that they repair throughout incubation because rotting vegetation at the bottom causes the nests to sink. Hawks, owls, herons, egrets, storks and other large birds that build nests in trees usually construct bulky structures of sticks and branches that may last for dozens of years. The birds merely repair damage or add new material to the lining each year. Most species that nest on cliffs have small, flimsy nests, but some, such as Bald Ibises, construct large, bulky ones. Songbirds' nests are usually small, sturdy and carefully woven.

← **Rosy Bee-eaters** look on as one of their number is busy digging its shallow nest in the sand.

↙ **In southern Africa, Social Weavers** build massive communal nests where dozens, or hundreds, of birds nest. They push and pull material into loops and knots in a technique that is similar to basket weaving. Each pair of birds has its own separate nest chamber.

↓ **Hummingbirds** often make their nest on a small branch or, as this Green Hermit has done, on a sturdy leaf. The nest is a small cup made from plant material and held together with spiderwebs.

→ **Fairy Terns** do not build nests. The single egg that the female lays is small enough to simply sit in a small indentation in a tree branch.

NESTING MATERIALS

Grass, leaves, weeds, sticks, branches, mud and the hair of mammals are just some of the building materials that birds use in their nests. In general, birds select materials that are readily available and suitable for the size and strength of the nest they need to build. Birds, especially those that nest in colonies, often steal nest material from neighbors rather than hunt for their own. When spray-painted nest material was placed in artificial nests in a colony of Franklin's Gulls, it soon became scattered right throughout the colony. Within a week much of it appeared in nests half a mile (0.8 km) from where it was placed. Some plants that birds add to their nests help combat ectoparasite infestations and infections. Feathers, snake skins, string, fishing line, pieces of plastic and glass all turn up in nests, presumably as decorations or deterrents.

Nesting behavior

Birds exhibit a great diversity of nesting patterns. Some are solitary nesters; a few species nest in huge colonies of hundreds of thousands of birds. Both patterns have advantages and shortcomings. Solitary nests can be hidden from predators; large colonies are clearly visible. When a predator approaches a solitary nest, its inhabitants usually creep from the nest and fly away, trusting that their camouflaged eggs and nest will escape detection. If the predator locates the nest, the resident birds may perform elaborate distraction displays to draw the predator away. Distraction displays are common in solitary ground-nesting species of shorebirds. Solitary nesting birds are also able to forage close to their nests, which reduces the time they need to travel to find food. Species that nest in dense colonies may need to fly some distance to feeding grounds. Nesters in colonies rely on neighbors to warn them of approaching predators, and employ such group defense strategies as direct attack or mobbing. The sheer number of birds also affords some protection by reducing the chances that an individual bird will fall victim to a predator. The amount of time that colonial nesters need to spend defending their territories puts them at some disadvantage. So, too, does the possibility that neighbors may plunder their nest materials, eggs or chicks. As a protective strategy against ground predators, both solitary and colonial species sometimes nest in trees or on cliffs.

↙ **White Storks are solitary nesters** that display a preference for building their nests on the roofs of houses or barns. The proximity of these nests to human dwellings has, perhaps, helped inspire the enduring fable that storks are the bringers of human babies. The fact that storks build elaborate and sturdy nests for their chicks, and that they use the same nests, year after year, may have helped to reinforce this myth. European Storks have been building their nests on chimneys and rooftops for so long that it has now become common for homeowners to adorn their roofs with structures that will attract nesting storks.

↓ **Hawks and owls** are both predatory species. They make solitary nests that help define these birds' territories and hunting spaces. Here, a Hawk Owl broods its chicks to protect them from the cold and predators. Female owls usually brood and feed chicks.

↑ **Although they are primarily ground predators**, Secretary Birds build large nests in trees in Africa. These solitary nests, made of sticks, leaves and grass, are usually within 20 feet (6 m) of the ground.

→ **Black-legged Kittiwakes** nest in large colonies on cliff ledges in Arctic regions. These birds often nest with other gull and murre species.

LIVING TOGETHER

Depending upon location, bird colonies may contain several different species. In temperate regions, mixed-species colonies may consist of gulls and terns, herons, egrets and ibises. In more northerly regions, murres, puffins, kittiwakes and cormorants may nest together. In many instances it is limited habitat, rather than preference, that drives diverse species to nest together. There are numerous examples, however, of different species nesting together in places where there is an abundance of available, unused habitat. Colonies of birds with mixed species bring a number of benefits. Large numbers of birds increase the likelihood that predators will be detected before they can strike. In mixed colonies, there is often less competition, both for space within the colony, and for nearby prey.

Inside the egg

During incubation a bird embryo develops into a fully formed chick that then hatches. A parent bird sits on the nest with the bare area on its belly, known as its incubation, or brood, patch, in contact with the eggs. This transfers the adult's body heat. Incubation normally lasts between one and twelve weeks. An egg's hard external shell must be strong enough to withstand the entire weight of the incubating adult. Inside this shell are albumen, or egg white, yolk and a small embryo. The yolk, which is an energy-rich food supply for the embryo, is made up of about 28 percent fats and 20 percent proteins; the rest is water. The yolk is housed within a yolk sac and functions early in the embryo's development as a stomach and intestines. Ultimately, the yolk is absorbed into the chick's body cavity. Albumen consists mainly of protein (about 10 percent) and water. It cushions the embryo from the shock of movement and insulates it from temperature changes outside the egg. The shell protects the egg contents from diseases and facilitates the exchange of respiratory gases between the embryo and the outside world.

↑ **Southern Giant Petrels** begin breeding on South Georgia Island when the snows are still falling. As they incubate, their brood patch shields the eggs from the wet snow, which is usually warmer than the surrounding air. In this species, male birds are responsible for the majority of incubation and guard duties.

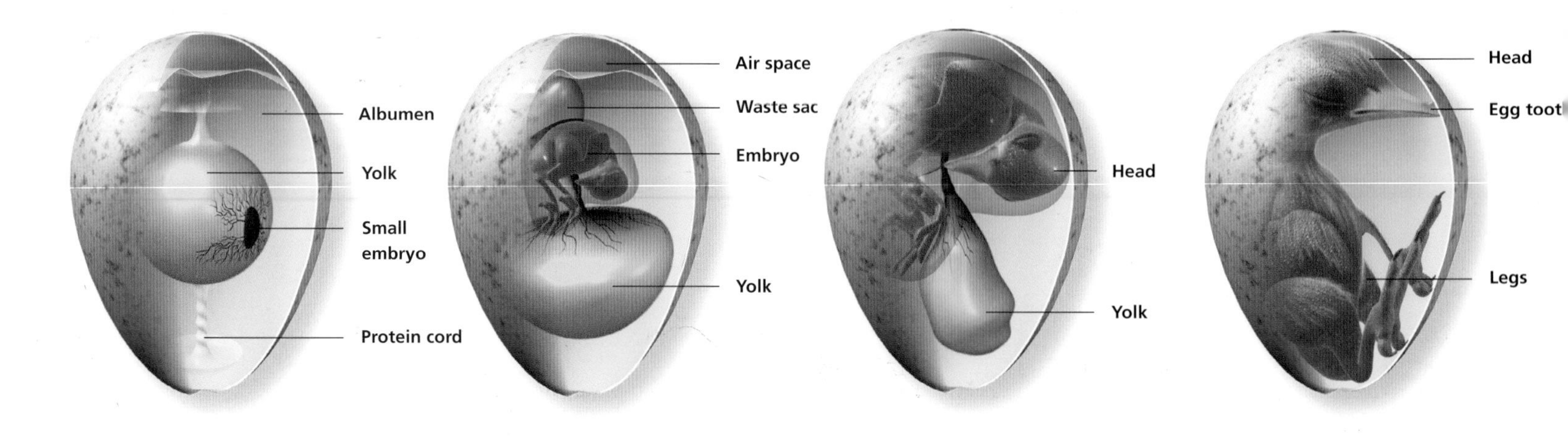

↓ **An egg contains all that is needed** for an embryo's growth. As it develops, using the yolk material, the embryo occupies a greater proportion of the egg's interior. There is gas exchange through the eggshell as the embryo breathes in oxygen and releases carbon dioxide. To hatch, an egg must be incubated.

HEARING A PARENT CALL

Although bird eggs have hard shells, the embryo begins its communication with the outside world right from the start. At first, this communication involves only the exchange of gases through the shell membranes. Later, as the embryo's sensory abilities develop, the chick is aware of sounds outside of the egg. Most of the calls that a developing chick can hear come from their parents, who call to each other during incubation exchanges or during pair-maintenance behavior. Experiments with Laughing Gulls have shown that young chicks that are prevented from hearing these sounds take longer, after they are hatched, to recognize the calls of their parents than chicks that were able to hear them inside the egg. This is critical because chicks are normally able to recognize their parents' calls when, about a week after hatching, they begin to walk around their territory. A chick at the edge of its territory that mistakes a neighbor's for a parent's call will probably be killed. Recognition of parental voices is particularly important for ground-nesting colonial species, where the chicks can wander.

↑ **In the tropics** of both the Old and New worlds, Greater Flamingos nest in dense colonies. They build large nests of soft mud, which soon hardens to provide protection for the eggs.

← **Common Eiders** line their nests with their own down to provide insulation for their large clutches. Only the females, which are dull colored for camouflage, incubate the eggs.

Hatching

The most traumatic time for a developing chick egg is at hatching, when it breaks the shell and emerges from it. Just before hatching, the chick stretches and fills the space inside the shell that was once occupied by the yolk and albumen. The chick first breaks the membrane inside the shell and saws with its egg tooth to make a small, starlike opening. Chicks alternate between sawing and pushing as they strive to get out of the shell. It is important that the chick is free of the shell before the membrane dries and attaches to its body. Otherwise, it may remain trapped inside, and die. Some "altricial" chicks hatch blind and naked; others are down-covered and able to walk when they hatch. As the natal down dries, it becomes soft and fluffy. Parents promptly remove the eggshells after hatching to protect the camouflage of the nest from aerial predators; the white inside of the eggshell is very visible from above. Eggs in a nest may hatch at the same time if all the eggs were incubated from the very beginning, but if incubation began at different times, they may hatch separately. Separate, or asynchronous, hatching results in newborn chicks of different sizes. In poor food years only the largest chicks will compete successfully for food and survive. Asynchronous hatching occurs in gulls, herons, egrets and raptors.

↑ **From the time a chick makes the first small hole** in the shell of its egg, it may take between 5 and more than 24 hours to hatch.

→ **These Canada Geese**, protected and guided by their parents, began to forage for food within hours of hatching.

BREAKING OUT

As they prepare to hatch, embryos develop a short, pointed egg tooth at the tip of the upper beak (shown at left on a parrot embryo). In most birds, this tooth falls off within a few days of hatching. To position the egg tooth against the inside of the shell, the embryo pulls itself backward and upward. It first breaks the membrane that encloses the air chamber at the blunt end of the egg. Then, while slowly rotating its body, the embryo pecks repeatedly at the shell until it has created a small star-shaped break. It saws at this opening until there is a large hole. A small muscle on the back of the neck, which atrophies once its task is done, provides power for this early pecking. Soon the stretching chick is able to break open the egg.

→ **A Mauritius Kestrel embryo** slowly breaks out of its egg. This species' hatching sequence is a fairly typical one. The chick begins the process by using its egg tooth on the tip of its bill, first to break the membrane and then to saw and peck the inside of the shell and smash a small opening in the hard outer shell. It continues to saw at the opening until the hole is larger, then it pushes with its body until the two parts of the shell are split open and it can emerge, struggling free. After hatching, young kestrels remain in the nest for about 38 days. During this time, they are completely dependent on their parents for food and protection. When they eventually leave the nest, they stay in their parents' territory for the first year.

Living in the nest

The way chicks develop between hatching and leaving the nest varies between species. However, birds exhibit two main patterns: altricial and precocial young. When altricial young hatch their eyes are closed, they have little down, they are immobile and entirely dependent on their parents. When precocial young hatch their eyes are open, they are covered with down, are able to walk almost immediately and have fairly good neuromuscular coordination. They require less parental care. The energy needed to produce eggs in altricial birds is low because these eggs are relatively small. Precocial birds, by contrast, invest heavily in egg production. In some shorebirds, the weight of the entire clutch is equal to half the weight of the female. Songbirds, woodpeckers and parrots are altricial, while shorebirds, quail, grouse and murrelets are precocial. Other birds, such as gulls and terns, are in-between. Hatchlings of these species can soon walk, but must be fed by parents. Until they fledge, altricial birds remain in the nest, and are fed by their parents. Precocial ducks, grouse and quail leave the nest almost immediately and, although they remain with their parents until fledging, forage on their own. In both altricial and precocial species, parent birds devote all their energies to protecting their young at this precarious stage.

→ **Brown Pelican parents** bring food back to the nest in their pouch. Chicks stick their bills down the parents' throats and pull out a half-digested meal.

↓ **Newborn Common Loon chicks** swim around with their parents, riding on their back when they are tired. At any hint of danger they hide under their parents' wings.

← **Great Gray Owls**, like many other species, carry prey back to their young in their bills. Owls can vary their breeding times and clutch sizes according to availability of food. A Great Gray Owl may not nest at all in years where food is particularly scarce, whereas in good years, it may lay up to nine eggs. Hatching is asynchronous, so that if food is hard to come by, the more advanced chicks can survive by eating their younger and weaker siblings. Young often leave the nest still partly in down.

FROM HATCHLING TO FLEDGING

The time between hatching and fledging varies from just over ten days for anis, to ten or more months for King Penguins. Songbirds take two weeks, kingfishers four weeks and hornbills seven weeks. Except for some slow-developing young, the length of the nestling period is related to the incubation period and to body size. The largest penguins (Kings and Emperors) brood their young for only about six weeks. After that, the young huddle close together with other young in large groups called crèches; their parents continue to feed them for up to ten months. The rate of growth and the safety of the nest also affect nestling periods. Ground-nesting birds in areas with mammalian predators leave the nest relatively quickly.

→ **Like all songbirds**, Golden Orioles are an altricial species. When they hatch they are blind, nearly naked and entirely dependent on their parents for food and protection. These chicks are being fed by their father.

Growing up

Fledging usually refers to leaving the nest. In a wide range of species, the chicks fledge when they are still unable to protect themselves from predators or forage entirely on their own. Birds abandon their nests when predators or weather conditions make it no longer safe for them to stay there. They remain with their parents while they learn to feed and avoid predators. Songbirds can fly, though weakly, when they fledge, but they still need parental care. It is critical for the young that they are able to recognize their parents, either by their calls, or by visual cues, or both. During this dependent period, most parents employ alarm calls to warn their young of danger. When chicks hear these calls, they run for cover or remain absolutely still, relying on their cryptic coloration to hide them from predators. Learning to feed is not routine. Young birds have to learn what foods are suitable for them and how to locate and capture them. The transition to complete independence is a difficult and dangerous one. In acquiring all the adult skills they need, young birds make mistakes, and these can be fatal.

↙ **Ostrich young leave the nest** as soon as they hatch, but travel with their parents for protection. When predators approach, adults warn their young and engage in a vigorous distraction display.

↘ **Emperor Penguin chicks** leave the nest at about six weeks old to join other chicks in a crèche. An adult penguin remains to guard the crèche. Parents continue to feed their own chick for eight or nine months.

→ **Harpy Eagle chicks** remain near the nest until they fledge. When close to fledging, chicks exercise their flight muscles by flapping their wings and jumping. This "practice" flying also strengthens their talons.

CONTINUING CARE

In a number of species of seabirds, parents continue to provide care even after their young have left the nest, learned to fly and achieved some degree of independence. Young gulls and boobies, for example, still return to the nest to feed after they have learnt to fly and explore. This period of post-fledging care gives young birds time to hone complex foraging skills and improve their flight maneuverability. In some colonial species, groups of young remain near breeding colonies in crèches, exploring their environment and practicing flying and foraging. In other species, such as some terns and pelicans, the young remain with the parents for many months after they leave nesting colonies. Gradually they catch more and more of their own fish. Great Frigatebird young are still partially reliant on parents as long as 400 days after leaving the nest.

← **Like most chicks**, Black-crowned Night Heron chicks spend their time in the nest until fledging, but they jump from the nest when their parents give an alarm call. They run wildly through the underbrush; dull, juvenile plumage helps to hide them. The chicks return only when parents land on the nest and give an all-clear call.

Lifecycles and life spans

Once a chick hatches, it passes through a period of parental care on its way to independence and adulthood. Adult birds have a reproductive cycle that includes finding a territory, and then a mate; building a nest; laying and incubating eggs; brooding and caring for young until they are independent; and surviving to the next breeding season. Most species have an annual reproductive cycle, although the timing of different stages in a bird's lifecycle varies between the different species. In some large birds, such as some albatrosses and King Penguins, more than a year elapses between when eggs are laid and the young achieve independence. The breeding cycles of songbirds are relatively brief. Most of them take only a week or two to find a mate; four or five days to build a nest; five to seven days to lay eggs, depending on the clutch size; and ten to fifteen days for incubation. Their young become fledglings in between ten and twenty–one days. For most other birds, each phase of the cycle takes much more time and energy.

↓ **The annual cycle of events** in most birds' lives is set, but the time required to complete each phase varies from species to species. Depending on species, events occur and phases begin in different months.

LIFECYCLE OF THE YELLOW WARBLER

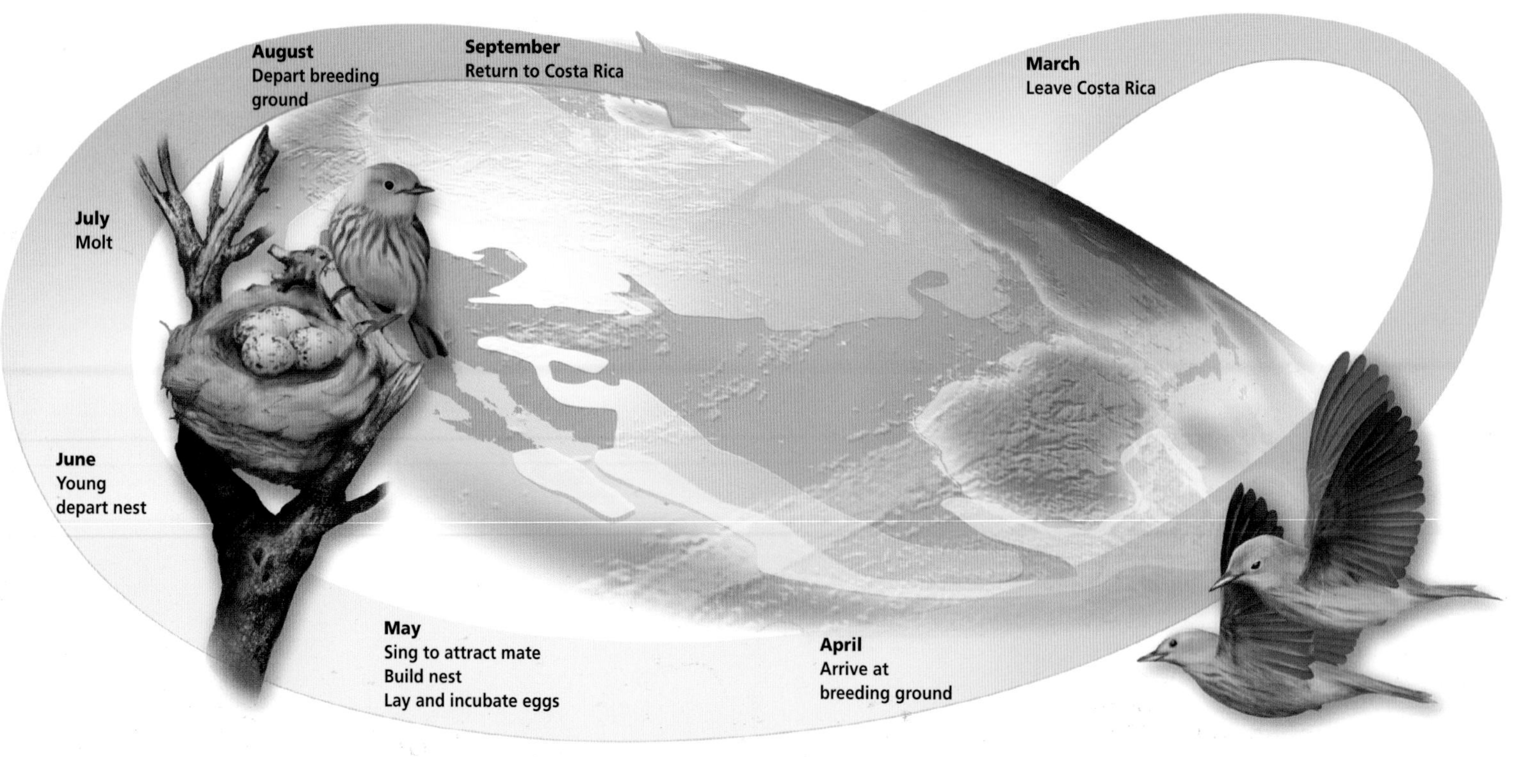

↑ **Songbirds, such as this Bluethroat** in Europe, have relatively short life spans and brief breeding cycles.

↖ **Galahs are endemic to Australia** and, as members of the parrot family, they enjoy a long life.

→ **Black-browed Albatrosses** are slow to find mates and their breeding cycle can take more than a year.

LONG AND SHORT LIVES

The annual survival rates of adult bird species vary from about 30 percent in small birds, such as Blue Tits and Song Sparrows, to more than 90 percent in penguins and albatrosses. These differences are reflected in different life spans. Most small birds with large clutch sizes and short parental care periods live short lives; large birds with small clutch sizes and long parental care periods live longer. Small birds such as warblers and sparrows live for between 5 and 10 years; some species of large birds can live more than 50 years. Japanese Quails, which live from 2 to 5 years, have one of the shortest life spans. Even within related groups, life spans vary. In the wild, American Kestrels live 13 to 14 years, while Bald Eagles can live to 25 years. Parrots have the longest life spans of any bird group. African Gray Parrots live between 60 and 70 years and some macaws survive to the age of 90 years.

Solitary and gregarious lifestyles

In their degree of sociability, birds range from solitary to highly gregarious. But they show varying degrees of sociability in the different behavioral areas of courting, nesting, roosting, migrating and feeding. Some birds, including blackbirds and grackles, herons and egrets, and gulls and terns, do nearly everything in groups. Others, including tinamous, loons, and some hawks and eagles, operate mainly alone or in pairs. Even within the same species there are variations in patterns. Great Blue Herons and Gray Herons, which often nest solitarily, at other times nest in colonies of several to a hundred pairs. Common Terns only rarely nest alone, but nesting groups can range from a few to thousands of pairs. Birds become gregarious when flocking increases their chances of survival, and facilitates foraging and breeding. The advantages of group living include early warning of predators, more effective group defense, enhanced ability to find mates and increased foraging efficiency. The old adage that there is "safety in numbers" is particularly true for birds; one individual in a hundred foraging birds has much less chance of being eaten by a predator than does a bird foraging alone or in a pair.

↑ **Cape Gannets** nest in dense colonies, where the edges of their nests nearly touch one another. They often feed in groups above schools of fish.

GROUP AND SOLITARY FEEDERS

Birds eat to say alive, and they forage in ways that enhance their chances of success and minimize their likelihood of being eaten. Whether a bird forages alone or in a group depends upon its preferred food, its hunting method and the dispersion and behavior of its prey. When food is abundant, and communal feeding does not curtail how much they can eat, some birds feed together. This is especially true when feeding with others allows them to catch more prey. White Pelicans forage in groups that drive and herd together large numbers of fish, making them easier to catch. The birds take turns being at the leading edge. Terns and boobies feed in dense flocks over schools of fish. The schools come together for only a short time and the number of birds does not affect what any individual can catch. Shorebirds feed in dense flocks in mudflats or along tide lines where the quantity of prey is constant. For seed-eating birds that forage on the ground, such as blackbirds, weaverbirds and grackles, the sheer number of feeders increases their security. Birds that feed on solitary prey, such as hawks, owls and some songbirds, usually feed alone.

↑ **Goshawks are solitary breeders** and they migrate alone or, but only rarely, in small groups. Like all predators on small birds, Goshawks are solitary hunters. This Goshawk has just captured a magpie.

← **Scarlet Ibis are social** most of the time. They breed in dense colonies, and in the non-breeding season they usually roost together at night. They also feed in groups, although feeding flocks are smaller than their breeding ones.

↞ **"Rookery" has come to mean** any large assemblage of breeding birds. It can even refer to breeding seals. It derives from the nesting habits of Rooks, which come together in large colonies throughout Eurasia. Each Rook builds a substantial nest made of twigs and small branches.

Display behavior

Birds communicate visually through display activity. A display is any color pattern or physical action—or any combination of these—that sends a message to another individual. Birds use display to attract and court mates, to warn off mating rivals, to communicate with their young, to defend nests and offspring, to establish and maintain territories and to announce danger. Although plumage color and elaboration are central to much display behavior in birds, there are often complex displays in which color plays no part. Cranes exhibit some of the most spectacular mating displays. These involve intricate and sensual dances, in which both partners jump up and down, wave their wings, dip and bow their heads, then lift their bills skyward, all the time calling loudly. In their courtship displays, boobies gently touch each other as they point their bills upward. Gulls and terns croon softly and raise and lower their heads as they circle each other in a dance. Some male songbirds scrape and bow on branches to show off their bright colors to a waiting female. Manakins, cocks-of-the-rock and birds-of-paradise create elaborate display areas. In some species, such as Ruffs, sage grouse and cocks-of-the-rock, many males display together on a lek. Females visit the lek and select their mates from the displaying males. To signal that it is about to take to the air, a bird may raise its wings slightly. To show aggression, it may raise the feathers on its head. Other clearly aggressive displays involve lunging the head and neck, raising the wings or moving menacingly toward an adversary. Wings and bills usually feature prominently in threat displays.

↑ **Sunbitterns in Costa Rica** have an elaborate threat display. They raise their wings to expose a strongly contrasted plumage pattern and vibrant eyespots that frighten off most predators or intruders.

← **Many bowerbirds** lack brightly colored plumage. To display, most build extensive bowers and decorate them with a variety of materials. Here, a Great Bowerbird is displaying to a prospective female at his bower. After mating, females leave to build their own nests.

MEANS OF ATTRACTION

Males display predominantly in order to attract mates. They use color and feathers as ornaments to attract the most desirable females. Male peacocks, birds-of-paradise and whydah finches have highly exaggerated tails, whose only function seems to be to attract females. In an elaborate display ritual, a male of these species fans or swishes its tail and body in order to get a female's attention. The King of Saxony Bird-of-paradise has two long head plumes that are decorated with a sequence of glossy blue, rectangular ornaments. It sings a piping, insect-like song as it waves these head plumes around in its mating display. Some male bowerbirds are quite drab-colored, but they display to females by building bowers, which vary greatly in their construction and complexity. Some bowerbirds simply sweep clearings in the forest floor. Spotted Bowerbirds select a bower and modify it only slightly by strewing leaves in front. The Crestless Gardener, also known as a maypole builder, makes an extensive, tentlike structure that resembles a thatched roof. He then clears a space in front, where he plants mosses and scatters brightly colored flowers. Satin Bowerbirds are the most distinctive decorators. Males mix charcoal, flower pigments and saliva to make a paint with which they adorn the bower walls.

← **Male Ruffs display to females** by expanding their ruffs as they dance on their lek. Males with darker ruffs usually predominate and have more matings with females.

Singing and calling

Throughout the ages, humans have enjoyed the sounds of birdsong. But birds sing only for each other. Bird vocalizations, performed by both sexes, range from simple clicks and calls to elaborate songs. The syrinx, a specialized organ that only birds possess, produces these sounds. It is located at the base of the trachea, or windpipe. All bird species give calls, which are simple sounds that serve a range of social interactions. Some birds, such as Mute Swans and Turkey Vultures, can only hiss and grunt. Most songs function to attract mates or to interact with other species. Mockingbirds and parrots can mimic other bird and non-avian sounds and incorporate these into their song repertoire. Some species learn their songs; some are born with them; most develop their innate, or acquired, abilities over time. Some species, such as snipe and woodcock, produce their loud, distinctive sounds by plummeting from great heights. Others, such as prairie-chickens and some grouse, make noisy, booming sounds by inflating air sacs at their throats. Woodpeckers strike hollow trees to create rich staccato calls. They have even learned to amplify their sounds by drumming on TV antennas or other metal objects.

↑ **To study bird songs**, scientists make sonograms which show differences in pitch, sound quality and length. Sonograms are like signatures for different species. In general, a straight horizontal line indicates a pure whistle at a particular pitch; a click appears as a vertical line. This Song Sparrow sonogram shows introductory whistled notes followed by a complex set of warbling syllables.

↓ **A Ruffed Grouse** produces a characteristic drumming sound by standing on a log and beating its wings vigorously. Grouse choose hollow logs to amplify their sound.

→ **European Robins** select one or more song perches that they use for their territorial or courtship singing.

↓ **Songbirds (oscines)** have up to six pairs of intrinsic muscles that operate the syrinx, which accounts for their elaborate songs. During normal breathing, the syrinx passageways are open. To produce song, the air column vibrates and the muscles contract, changing the diameter of the passageways, and putting tension on the typaniform membranes.

INNATE AND LEARNED SKILLS

Natural selection has molded the development of bird songs, and the role of learning in their singing. At one end of the continuum, White-crowned Sparrows depend entirely on their parents to learn their distinctive song. Gray Catbirds, on the other hand, seem to need no parental tuition to sing the song that identifies their group. Catbirds that are isolated from all other birds can sing an amazing variety of songs. If given the opportunity, they can learn many more. Young cuckoos, raised in the nests of other species, never hear the song of their parents while they are developing into adults. Once they reach adulthood, however, they make typical cuckoo sounds. Many factors affect a bird's song development: the sound of its own voice stimulates learning; the songs of parents and neighbors enhance the quality and range of its song; and hormone levels affect vocal quality. During the breeding season, the levels of testosterone decrease in males. As a result, these birds become less melodious. Eventually they stop singing.

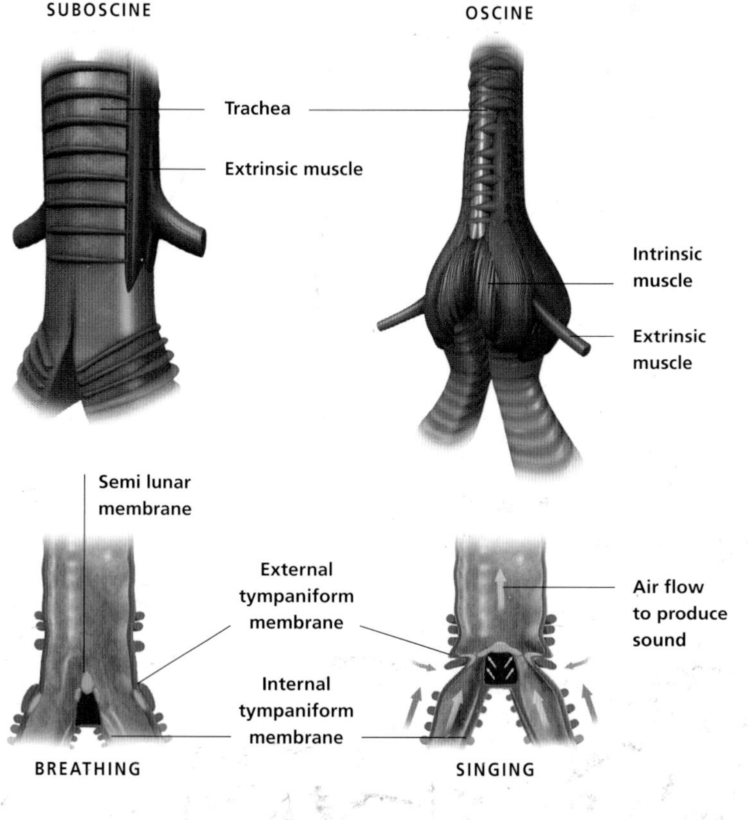

On the attack

When birds fight, they do so for the purposes of defense or acquisition. A bird will attack another to defend itself, its mate, its offspring or its nest from a perceived threat. It may also use aggression to catch prey, to acquire nesting materials or to claim a nesting site. The primary object of an overt attack is often a member of the same species, but other species may also be targeted if they are seen as territorial competitors or unwanted intruders. Starlings, for example, evict other birds from nesting holes. Some species, such as Australian Magpies, will swoop and snap at people who, however innocently, come within the vicinity of their nests. Any intruder, avian or otherwise, that enters the space of foraging or nesting birds may represent a threat. When a bird launches an attack it is usually for a specific purpose; unlike some mammals, birds do not attack and kill wantonly. Often a threatening display will be enough to frighten off an offender, and so make overt attack unnecessary.

LIMITS TO AGGRESSION

Birds can be aggressive only within the limits imposed by the habitats they occupy. Birds that nest on small cliff ledges, such as kittiwakes, are non-aggressive. This is because aggressive behavior could send adults, chicks and eggs tumbling off the ledge. But kittiwakes are also relatively safe from attackers because of the small size of their nesting places. Birds that nest in holes in trees are not aggressive near their nest sites, because the vertical trunks of trees do not provide adequate sparring places. But where nest holes are in short supply, fights may break out. On the other hand, birds that nest on the ground or on other flat surfaces have sufficient space to attack one another. Many ground-nesters in mixed-species colonies have prolonged and frequent fights with neighbors and predators. Nesting on the ground also exposes birds to a wide range of unintentional intruders that only an overt attack will repel. Common Terns need to be alert for diamondback terrapins, which crush terns' eggs as they seek out their own nests. Horses, cows and sheep can also endanger eggs in ground nests, as well as chicks.

↟ **Atlantic Puffins fight**, often on steep slopes or rocky ledges, by pulling at each other's bills while flailing their wings. One bird eventually loses its balance and, being the loser, must fly away.

↑ **Young fulmars defend themselves** by spitting a foul-smelling, oily vomit directly at any approaching predator. Fulmar means "foul gull" in Norse.

← **A Bullfinch (far left) threatens a Tree Sparrow.** Bullfinches must compete with other small species for food and nesting sites.

→ **Martial Eagles** prey on young warthogs. Here, one threatens a female warthog as she defends her young.

Defense and diversion

Blending in with the background is an effective defense against most intruders and predators. If a bird is not readily visible, it can escape capture. Birds that are resting, nesting or roosting on the ground, and remain immobile, rely heavily on camouflage for protection. Ground-nesting birds often have cryptically colored nests, eggs and chicks that closely match the colors of their surroundings. When a predator or intruder approaches, the parent may skulk away, taking to the air only when it has moved some distance from the nest. When the predator sees the bird fly, it runs to the take-off point, thinking that the nest and young are there. Birds that depend on camouflage need to be able to signal to their chicks to stay still. These species all have alarm calls that instantly immobilize the chicks. Camouflaged chicks remain quiet and do not vocalize until their parents have uttered an all-clear call. Some ground-nesting adults are also cryptically colored. These birds will usually wait until an intruder is almost upon them, then burst from the nest explosively, startling the intruder into departing. Some leave their departure so late that it is possible to catch them by hand. Most birds that build their nests on the ground rely on camouflage, but some nightjars are also colored gray to blend in with the tree branches on which they roost during the day.

DIVERSIONARY TACTICS

Many ground-nesting birds employ broken-wing displays as a form of defense. A bird flopping on the ground will always attract a predator. The displaying bird calls out in feigned alarm, and the predator follows it, seemingly assured of a meal. This display distracts the intruder from any interest in the nest or chicks. When the bird has lured the predator to a safe distance, it suddenly "recovers" and flies away. The other main type of distraction display is the “rodent run.” Shorebirds will often distract a predator away from the nest, then run rapidly in a low crouch that appeals to the mouse-catching instincts of mammalian predators. Once the danger is passed, the bird sneaks back to the nest. Scientists have argued for many years about whether distraction displays are a sign of intelligence or simply an automatic behavioral response to a threatening stimulus. Intelligence seems the more likely explanation. The distracting parent always carefully monitors the predator's movements and alters its own behavior accordingly. It affects subtly differing types of injury and gauges the speed it needs to remain clear of the intruder.

← **Stone curlews rely on camouflage** for defense and steal away from the nest when threatened. If surprised, however, they perform a distraction display.

← **Pauraque Nightjars** have perfect camouflage for the forest floor. Their plumage has a wide range of browns, blacks and beiges that matches the patterns of leaves and twigs.

→ **The irregular stripes** on a Great Bittern's throat and chest blend in with surrounding brown reeds as it points its bill toward the sky.

↓ **Skimmer chicks**, covered in white down, are almost invisible against the pale sand in which this species scrapes out its flimsy nests.

Territorial behavior

Birds turn aggressive with the onset of the breeding season as they vigorously stake out and defend claims for space. Territorial behavior is not limited to breeding; birds will also defend feeding and wintering territories. We can define a territory as any defended space. Territory owners usually defend their spaces with threat displays; if these do not work they will resort to physical attack. Optimally, birds want exclusive use of their territories. Territorial defense is usually directed toward members of the same species who might usurp space, steal mates or compete for food. Species that nest in mixed-species breeding colonies must also defend their space against other species that could compete for the same nest sites. In some cases, a bird may defend a foraging territory for only a short time. When the amount of food obtained does not warrant the time and effort involved in defending it, a bird will move on. Bigger birds generally defend larger territories. The needs of developing chicks are also important; species whose chicks wander away from the nest usually have larger territories. For species that forage for food entirely within their breeding territories, the area must be sufficient to sustain both adults and chicks, whose energy demands increase greatly as they grow, throughout the nesting season. The size of a breeding territory will often change in response to changing conditions. When lemming populations in the Arctic are high and food is plentiful, Pomarine Jaegers defend smaller territories than when food is less abundant. Sunbird and hummingbird territory sizes vary with the number of nectar-producing flowers in the vicinity. In most parts of the world, birds begin to claim territories in the spring. Once they have established a space, they defend it for as long as is necessary.

TERRITORIAL PATTERNS

Many songbird species defend territories that provide all their food and nesting requirements. This increases feeding efficiency as the birds are able to forage for food near their nests. In most songbird habitats, birds are well spaced out, with breeding males singing every few hundred feet. Some species, such as blackbirds, grackles and barbets, get most of their food from their territory, but nest in loose colonies that require them to feed elsewhere as well. Several families nest in dense colonies, where they defend only enough space to incubate their eggs and raise their young. These birds have to leave the colony for food. Gulls, terns, herons, egrets, cormorants, puffins and many other seabirds normally feed within the vicinity of the nesting colony. Other seabirds, however, need to travel long distances to their feeding grounds. Some albatrosses may be absent for days, as they fly hundreds of miles to get food for their chicks. Penguins often walk many miles over ice floes, then swim even further, to find food. Many species return to the same territory year after year, and there is great competition for the best sites. Gannets compete keenly for their nesting sites, which they continue to defend for several months after their chicks have fledged.

Ring-necked Pheasant males often engage in territorial clashes that involve wing flailing and pecking. They are native to Asia, but varieties have been introduced into many countries—they are highly desired game birds. Females are drab and dull colored. They go off to incubate on their own and do not defend territories.

Coots are promiscuous and males must fight for females. Here, American Coots flap and splash frantically, emitting a series of loud whistles and grunts. They steam toward one another, rearing up just before they collide. Females usually remain passive while males engage in territorial disputes.

Preferring a threat display to overt attack, a Rockhopper Penguin stands on its nest, flings out its wings, erects its vibrant yellow crest and gives a honking scream to ward off intruders. These birds stand only 2 feet (60 cm) tall.

Even domestic birds defend their space, mates or chicks from others. Here, a domestic White Goose gives a classic threat display, extending and twisting its neck to threaten an intruder. It usually accompanies this visual display with a sharp snakelike hissing sound.

Grooming and hygiene

All birds care for their bodies, and for all species daily maintenance of feathers is critical. Because feathers are inert and do not have internal nourishment, they can become brittle with age. Birds condition their feathers with a waxy secretion from the uropygial gland, which is located at the base of the tail, on the rump. This oily secretion, which may also repel feather lice, is made up of waxes, fatty acids, fat and water. As it preens, the bird applies the oil to its feathers with its bill. Birds also preen in order to rearrange their plumage and reposition out-of-place feathers. To maintain the condition of the vanes they pull their feathers through their bills. During preening, a bird also breaks and dislodges the sheath that covers new feathers as they grow. Sometimes birds preen as often as once every hour. As well as feathers, bills, legs and feet require regular attention. The outer sheath of the bill (called the rhamphotheca) is constantly worn away as a bird wipes and scrapes its bill during feeding. Most birds clean their bills by wiping them on the ground, or on branches or other objects. Parrots wipe their bills with their claws or rub them against hard surfaces. Ducks and waterbirds clean their bills while they are bathing. Birds far from water, such as House Sparrows, Wild Turkeys, Greater Rheas and Ostriches, all dust-bathe with abandon.

→ **Like nearly all birds that bathe in freshwater**, Roseate Spoonbills dip their bills and bodies in the water and flap their wings strongly to shower themselves with water. After bathing, they find a safe place to preen their feathers and dry off.

↘ **This falcon, a Eurasian Hobby**, is scratching its bill and face with its claws. Most birds use their feet to scratch and preen feathers on their heads and other hard-to-reach places.

↓ **Anhingas lack the oil** that other birds apply to their feathers in order to waterproof them. This allows them to dive deeper and to move through water more efficiently. In return, their plumage becomes waterlogged, so after fishing, swimming or bathing, they stretch out their wings and dry them in the sun.

↑ **A Quaker Parrot**, also known as a Monk Parakeet, carefully pulls each feather through its bill to maintain the condition of the vane and reconnect the barbules.

↓ **Birds that lack access to water**, such as this Gray Partridge, often bathe in sand, flapping their wings and tails up and down to thoroughly douse their feathers.

BATHING TECHNIQUES

To clean their feathers, bills and feet, birds bathe in water, snow or dust. Bathing methods are consistent within the different bird families. Swallows dip down to the water's surface several times to wet their feathers. Terns plunge into the water from above and wash themselves vigorously. Ducks, gulls and other aquatic birds bathe by dipping their heads and bodies under the water, then rocking back and forth until they are almost completely wet. Most songbirds stand or squat in shallow pools or puddles, rapidly ruffling their feathers to splash water over their bodies. Many forest birds dip their wings delicately into small pools in trees or leaves. Some parrots stand out in tropical downpours, flapping their wings and turning upside-down to get their underbellies wet. After bathing, birds always preen to clean and rearrange their feathers. Dust-bathing is common in species that spend a lot of time on the ground, or that live where there is little water.

Finding food

Birds exhibit great ingenuity, and expend considerable time and energy, in their hunting and foraging. The number of strategies they employ and the range of foods they eat are as various as the number and kind of habitats in which they live. Different birds hunt at different times of the day and night, in all kinds of environments, eating foods as diverse as apples and zooplankton. Some birds forage alone; others do so in association with members of their own or other species. A few birds forage simply by piracy, stealing food that other birds have collected. Diet, habitat and the availability of suitable food strongly influence foraging methods and behavior. So, too, does a bird's anatomy. The shape, size and strength of a bird's bill, feet and wings largely determine the kinds of food that it can find and eat. The size and structure of a bill can tell us a good deal about a bird's foraging behavior. Birds with thick, heavy, hooked bills use them to tear prey; long, pointed, heavy bills are for spearing fish or amphibians; short, thick bills can crack open seeds; and small, thin bills are adapted to picking insects off trunks or leaves. Much of the food birds eat is seasonal; almost none is available all year round. Most species, therefore, must adapt their diets and their hunting and foraging strategies to some extent at different times of the year.

↟ **Woodpecker Finches**, from the Galápagos Islands, select exactly the right-sized cactus spines to extract insect larvae and grubs from holes in dead wood.

↑ **Cattle Egrests** feed on insects that have been scared out of the deep aquatic vegetation by a pair of grazing hippopotamus.

← **Great Blue Herons** wait silently and patiently, often for hours, until they seize suddenly on their prey. Most herons have long bills to stab prey.

EXPLOITATIVE FORAGING

Some species use beaters to scare up prey. Cattle Egrets on the African plains are masters of this art. They run after wildebeest, zebras, impalas and elephants, swooping on the grasshoppers and other insects that the animals' movement frightens into the open. In other parts of the world they follow cattle, horses and even tractors. Franklin's Gulls in the United States, and Black-headed Gulls in Europe, also follow tractors to pick up grubs and insects from the freshly overturned soil. Cowbirds and oxpeckers ride around on animals, taking ticks and other insects from their backs, bellies and legs. In the New World tropics, many small birds, including White-plumed Antbirds, Bicolored Antbirds, antshrikes and Black-crowned Antpittas, feed beside swarms of army ants on insects that the ants disturb. Ocellated Antbirds forage at the leading edge of the ant swarm, where the concentration of prey is highest. Other birds, such as Spotted Antbirds and Gray-headed Tanagers, follow the ants but forage higher in the trees. Drongos and hawks in grassland areas specialize in flying at the leading edge of grass fires, catching insects as they flee from the flames.

A Great Gray Owl swoops silently toward an unsuspecting mouse. The feathers on its feet and legs protect it against the cold and its hearing is so keen that it can detect rodents even beneath the snow. Adaptations such as these allow Great Gray Owls to remain and hunt for food in the Northern Hemisphere's boreal forests throughout winter.

Flowers, fruits, nuts and seeds

Birds consume a range of plant material. Species as diverse as Wild Turkeys, seedeaters, grass-quits and Zebra Finches feed on grass seeds; Mistletoebirds, cotingas, toucans and most of the tanagers specialize on fruits; honeycreepers and sunbirds forage on nectar. Fruits are rich in carbohydrates, vitamins and, in some cases, lipids. Acorns and other nuts are rich in fats and proteins. Ducks and geese often eat algae in addition to aquatic vegetation, and grouse feed on over 350 different species of plants. Many bird species that eat predominantly seeds, fruits and nuts, feed insects to their young because these provide protein and energy and are more easily digested. Even in tropical regions where the climate is mild, birds change foods seasonally because their preferred fruits or nuts are not always available. There is a symbiotic relationship between some fruit- and seed-eating birds and the plants they feed on. Birds derive nutrition from seeds, fruits and nuts and they recycle seeds when they defecate. Some seeds germinate more rapidly once they have gone through a bird's digestive tract. Indeed, the seeds of some Australian trees will germinate only after they have passed through the gut of a cassowary.

→ **A female Red-knobbed Hornbill**, perched in a tree, feeds on a small fruit. Hornbills play a significant role in the dispersal of seeds in tropical rain forests. They eat mainly fruit, and in particular feed on figs.

→→ **Black-throated Jays** feed mainly on nuts. They cache the nuts they do not eat immediately or feed to their chicks for use during the winter. Jays are great food-storers: Blue Jays store acorns, hazelnuts and sunflower seeds; Steller's Jays bury acorns; and Scrub Jays hide an array of seeds.

↓ **During the winter** birds often feed on inferior fruits that they do not need to eat during the breeding season. Here, a blackbird feeds on crab apples.

FOOD IN STORAGE

Some birds hide seeds and nuts for later consumption, though much of their store may disappear. Red-breasted Nuthatches, Eurasian Nuthatches and White-breasted Nuthatches hide up to 40 seeds in an hour in tree bark, only to have them stolen by creepers and chickadees. Some birds remember where they have stashed their seeds; others forget. Some birds bury their seeds in the ground and in fallen logs. Eurasian Thick-billed Nutcrackers relocate up to 70 percent of their stored pine seeds, even when they are covered with snow in the middle of winter. Crested Tits in Europe obtain nearly 60 percent of their energy in the winter from seed caches. Shrikes store larders of insects, mice and small birds that they hang on tree crotches or bushes, or on poles and fences. They sometimes return up to eight or nine months later to eat the air-dried prey. American Kestrels also cache the remains of their prey. Although most birds cache food for the winter, some species return for the food in the spring to feed their young.

→ **Because they can hover** by individual blossoms, hummingbirds are uniquely adapted to obtain nectar from flowers.

↓ **Hawfinches have thick bills** and powerful jaw muscles that they use to crack open hard seeds. They husk the seeds by manipulating them with their tongues.

Insects and herptiles

Insects, amphibians and reptiles are important in the diet of many birds. For most birds in temperate climates, insects are the most valuable animal food, especially during the breeding season. In stomach analyses of 80,000 North American birds, insects made up 88 percent of the items found. Most passerines, which comprise more than half of all bird species, are insect-eaters. In some predominantly seed-eating groups, such as grosbeaks, sparrows and finches, adults eat insects during the nesting season and feed them to their young. Young birds digest insects more easily than seeds. Grasshoppers, locusts, dragonflies, damselflies, bugs, butterflies, moths, beetles, flies, mosquitoes, ants and bees all fall prey to insect-eating birds. So, too, do the larvae of many insects. In their predatory foraging, birds select prey that provides maximum energy for its size. Amphibians and reptiles—together known as herptiles—are rich sources of energy but, as they are difficult to find and catch, they do not feature prominently in most birds' diets. Birds as diverse as herons, egrets, ducks, kingfishers, crows, grackles, hawks and some thrushes prey on amphibians. Fewer species eat reptiles; snakes and lizards are the major reptilian foods. Hawks, owls and crows are the most significant reptile-eaters, although some songbirds occasionally eat small snakes that they may mistake for worms. Cranes, herons, egrets, ibises and Wild Turkeys also eat reptiles when they encounter them. Among raptors, only the Secretary Bird feeds mainly, but not exclusively, on snakes.

→ **Flycatchers sit on branches** and sally forth to catch flying insects in midair. Here, a Brown-crested Flycatcher holds a recently caught grasshopper.

↓ **Red-backed Shrikes** eat a range of foods, including grasshoppers, butterflies and lizards. They impale some of their catch on thorns and branches to feed on later.

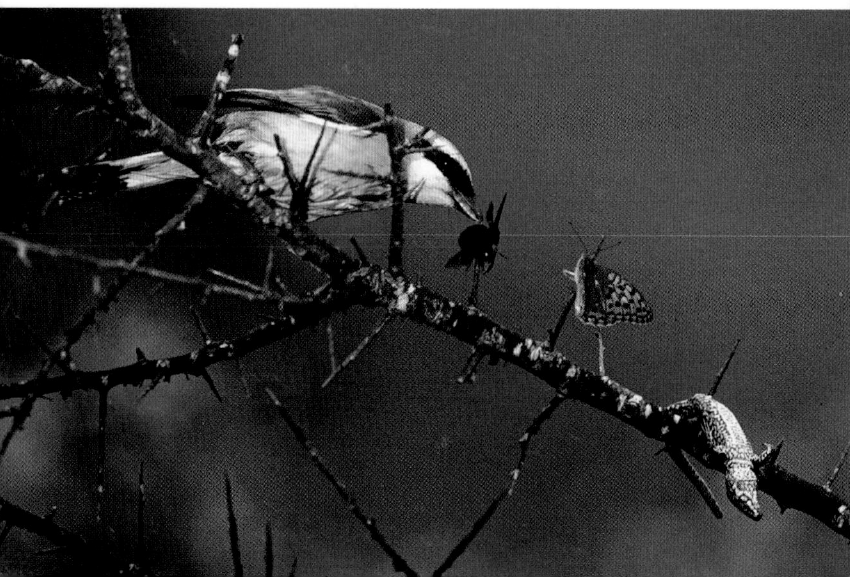

↑ **Short-toed Eagles swoop down** from high in the air to catch small reptiles, including snakes and lizards, that they often swallow whole.

← **Green Herons usually forage patiently** from small branches that hang over the water. Sometimes they wade through shallow water to catch frogs and toads.

PREYING ON SNAKES

Most snakes that birds catch are large enough to provide a lot of energy—sufficient to compensate for the energy a bird expends in seeking them out and capturing them. Hawks and eagles will sometimes kill snakes by dropping them from a great height before eating them or feeding them to offspring. Some snake-eaters, such as Red-tailed Hawks, quickly decapitate their catch. For some bird species a snake's venom is no deterrent. Bald and Golden eagles catch and kill rattlesnakes; they even carry them back to the nest to feed to their young. In rare cases, shrikes, too, prey on poisonous snakes. Many birds that do not specialize on snakes will catch one if the opportunity arises. But hunting snakes can be dangerous work. Hawks, including Red-tailed Hawks, sometimes fall victim to their prey's venom, and smaller birds that try to catch constrictor snakes are likely to perish in their coils.

Fish and seafood

Most seabirds eat fish or squid that they catch in estuaries or oceans. Many other species, including divers, grebes, herons, egrets, bitterns and some eagles, also feed mainly on fish. Some songbirds, such as catbirds and waterthrushes, occasionally catch small fish at the water's edge, and dippers sometimes eat fish and fish eggs in mountain streams. Most fish-eating birds eat fish all year, and they feed fish to their young. Adults generally eat larger fish than the ones they feed to chicks. Terns and puffins use their bills to deliver fish and shell-fish to their young; gulls bring the food in their gullets; albatrosses and cormorants carry it in their stomachs. Shearwaters reduce the fish, through their digestive systems, to an oil, which they then feed to their chicks. To provide a diversity of nutrients, most species bring their young a variety of fish types. Birds employ a range of strategies to forage for seafood: oystercatchers, gulls and shorebirds dig shellfish out of the sand; petrels, shearwaters and albatrosses pluck fish and squid from the water's surface; terns and boobies dive from great heights; and cormorants and auks swim underwater in their pursuit of prey.

LEARNING TO FISH
As they grow to adulthood, young birds that rely on fish need to learn not only how to catch their prey on the high seas, but where to locate it. The long and difficult training process accounts for the length of time that many young seabirds spend with their parents after they have fledged. In this post-fledging period, parents continue to provision their young, gradually allowing their offspring to assume greater responsibility for their food needs. Frigatebirds remain with their parents for between four months and more than a year after they begin to fly. These birds do fish for themselves, but they must also learn to pirate food from other seabirds, a skill that requires dexterity and maneuverability and that is acquired only gradually. Boobies and terns, both spectacular plunge-divers, receive post-fledging care for up to six months as they slowly master the difficulties and complexities of their foraging method.

↑ **Common Puffins carry fish** back to their offspring in their bills. Puffins' bills are brilliantly colored during the breeding season, after which the colors fade until the following year. The numerous ridges on a puffin's upper bill allow it to carry many fish at a time. It will bring food to its chicks until they can fend for themselves.

↗ **A Common Kingfisher plunge-dives** into the water when it spots prey. It then flies underneath the surface, snatches a fish in its bill, flies back to the surface and alights on a tree to eat it undisturbed.

→ **Glaucous-winged Gulls** exploit the prey of grizzly bears. The bears do the hard work, catching salmon that spawn in Alaskan streams, while the gulls wait in the wings to eat the remains. Gulls are voracious feeders.

← **Brown Pelicans plunge-dive** from well above the surface, opening their bills to scoop fish into their pouches. They catch fish mainly near the surface. Sometimes gulls and terns wait to steal fish from them. Other pelican species feed as they sit on the surface, dipping their bills and pouches down into the water. Groups may even herd fish into shallow water where they can be caught more easily.

Migration strategies

In order to survive, birds need to maintain their body temperature and obtain enough food and water for their daily energy needs. Most birds, including many tropical species, can achieve these goals by staying in one place. Those that cannot, migrate. Nearly half of all birds divide their time between two main locations. Every autumn millions of birds leave Europe, Asia and North America for warmer southern climes; they return in the northern spring. Migrating birds, unlike those that use torpor, migrate to exploit seasonal foraging opportunities throughout the year. Migration patterns are as varied as the migrating species. Birds migrate solitarily, or in single- or mixed-species groups. They set out at night, during the day, or at dusk. Some travel very short distances—perhaps just up or down a mountain; others move between the Earth's poles. Not all migrations are between north and south, and not all take place in spring and autumn. Although temperate birds migrate north and south to avoid cold weather and escape food constraints, many tropical birds migrate in response to rainfall variations and drought conditions that can affect the availability of fruits, nuts, seeds, insects and nectar. The type of migratory strategy that a species employs depends on what resources—such as food, water and shelter—it needs to successfully complete its journey. Migrating is a high-risk enterprise and only about half of all migrants reach their destinations. In contrast, between 80 and 90 percent of non-migratory tropical birds survive till the following spring. Perhaps to compensate for this, many migratory birds have higher reproductive rates than non-migratory species.

MIGRATORY PATTERNS

In their migratory habits, birds can be "residents," "nomads," "irruptive migrants" or "obligate migrants." The variability and reliability of food supplies are the keys to these categories. Residents, such as cardinals and woodpeckers, do not migrate because their habitats ensure adequate food supplies all year round. Nomads, such as budgerigars and crossbills, enjoy less reliability, but relatively low seasonal variation, in their food supplies. Nomadic birds will move when resources in one place become inadequate, but generally to a location similar to the one they have left. Irruptive migrants, which include owls and hawks, migrate irregularly. When food supplies dwindle in their regular habitats, they leave en masse for other regions. Obligate migrants must migrate each year from north temperate regions to warmer climates. Many thrushes and warblers belong to this group.

Many migrating birds suffer exhaustion and take refuge in trees and bushes where, like this migrating juvenile Magnolia Warbler, they rest and refuel before resuming their journey.

MIGRATION ALTITUDES

Bird Group	Average Height	Maximum Height
Plovers, sandpipers and other shorebirds	4500 ft (1500 m)	15,000 ft (5000 m)
Ducks and geese	2000 ft (750 m)	5500 ft (1700 m)
Swans	2000 ft (750 m)	10,500 ft (3500 m)
Hawks, vultures, cranes and storks	Thermals below 3000 ft (1000 m)	4500 ft (1500 m)
Hawks, vultures, cranes and storks over mountains	Depends on mountain height	26,900 ft (8205 m)
Songbirds over water	6000 ft (2000 m)	15,000 ft (5000 m)
Songbirds over land	2200 ft (750 m)	6000 ft (2000 m)

Different groups of birds migrate at different altitudes. The average height at which birds fly during migration is usually well below the maximum height they reach when winds are unfavorable at lower levels. Migrating birds may also fly at higher altitudes to surmount physical barriers such as mountains.

Snow Geese begin their annual southward journey between August and November. Huge migrating flocks fly at high altitudes and reach speeds of up to 50 miles per hour (80 km/h). They stop to refuel and recuperate at familiar places along the way. The geese eat copiously to build energy for their long flights.

Migration routes

Migrating birds follow the same route every year, using known mountains, rivers, coastlines, and the sun and stars, as signposts along the way. They also employ acoustic, magnetic or olfactory cues in their navigation. For young birds, their first migration can be a difficult one. Many follow their parents; others seem to rely on some innate sense of direction. Different species follow widely different routes. Golden Plovers and Hudsonian Godwits, for example, take advantage of prevailing seasonal winds. Waterfowl generally migrate above prominent waterways. Many species take indirect routes between their breeding and wintering grounds. In spring, shorebirds pass through central North America on their flight from their wintering grounds in the South American pampas to the high Arctic. In the autumn, they fly southeast to the Atlantic before turning toward the pampas. Hawks and some other birds often fly around lakes and bays to avoid long flights over water. The young of some species, such as shorebirds and hawks, migrate along coastlines, while the adults take a more inland route, exploiting mountain updrafts.

SITE LOYALTY

Birds develop a strong loyalty to certain places. They exhibit this "site loyalty" toward breeding territories, places along migration routes and wintering grounds. In banding studies in the United States and northern Africa, the same individual songbirds were caught year after year in exactly the same locations. Similarly, migratory birds are caught in the same woods, forests and grasslands in their southern wintering grounds. Hundreds, or thousands, of migrating shorebirds regularly visit such places as Delaware Bay in New Jersey, Texel in Holland and Copper River Delta in Alaska. Beaches have abundant food supplies for these species and individuals can almost double their weight during a two- to three-week beach stopover.

→ **Lesser Flamingos** of Africa's Rift Valley nest in colonies of up to a million pairs. These nomadic birds occupy one lake for months, then move on to another, often distant, one.

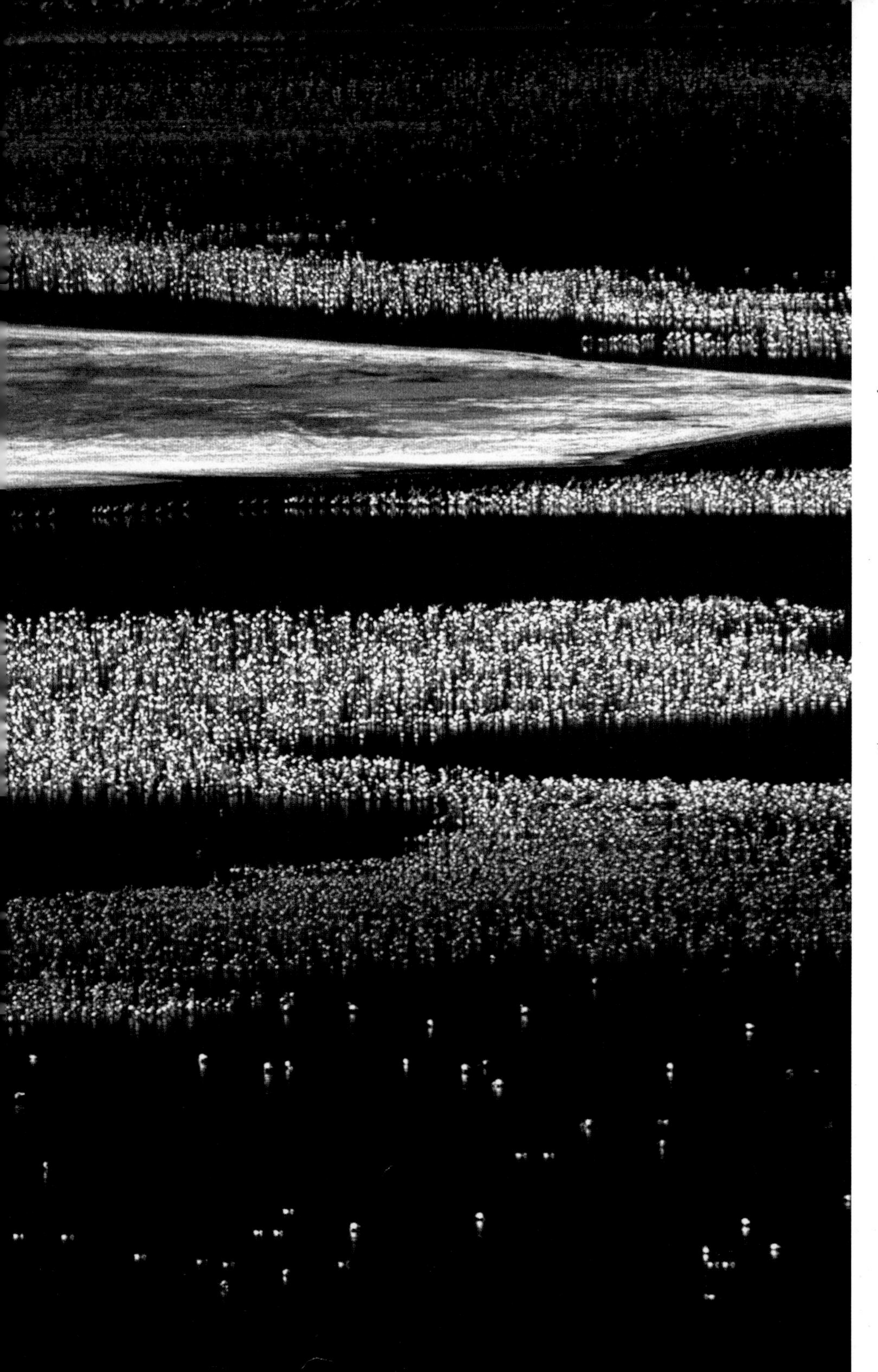

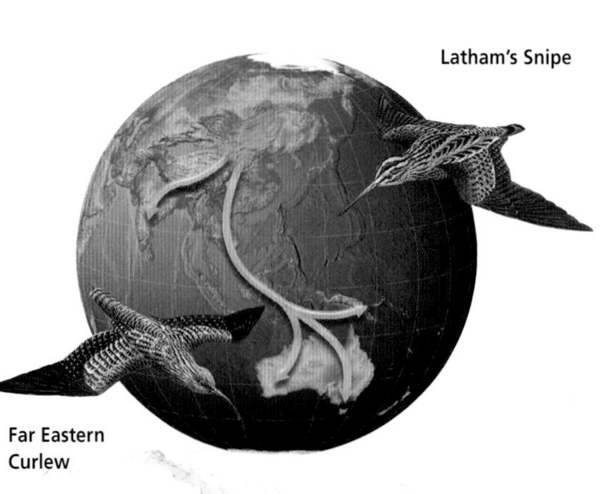

Latham's Snipe migrate only short distances; Far Eastern Curlews travel between continents.

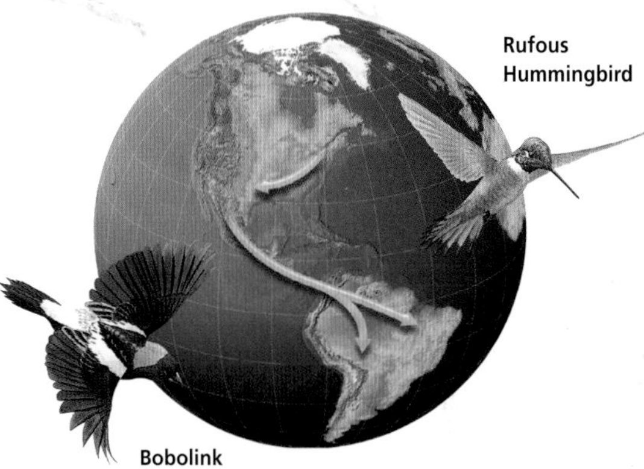

Rufous Hummingbirds move from Alaska to Mexico; Bobolinks migrate through Central to South America.

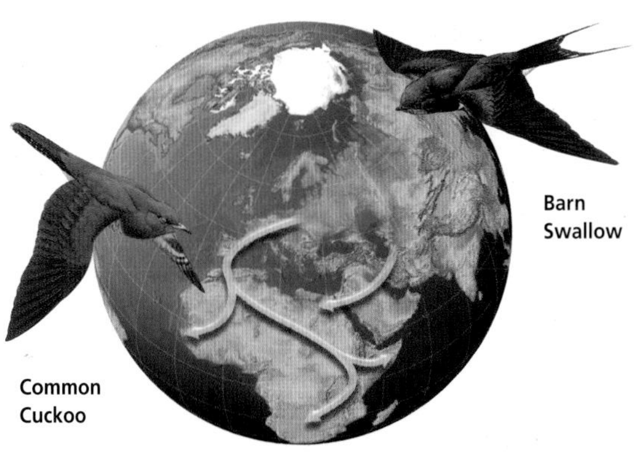

Barn Swallows and Common Cuckoos migrate along heavily used routes between Eurasia and Africa.

Kinds of birds

Kinds of birds

More than half of all known species are passerines, or perching birds. Other groups exhibit a diversity of form and behavior. On the following pages, the distribution of each bird group is shown in purple on a world map.

Naming and classifying birds

For centuries, humans have been naming birds and trying to put them into logical groups. Birds' names varied from one geographic area to the next until the mid-1700s, when Carl Linnaeus devised a system for naming each and every organism. This became the universal standard and is still in use. Each species is given a two-part Latinized name made up of its genus and species. Whereas common names vary from language to language or place to place, scientific names generally remain constant.

Finding a universal system for classifying birds has proved more difficult than agreeing on a naming convention. Early humans divided birds into two groups, those that were edible and those that were not. In the 1600s, Ray and Willughby classified birds based on similar morphological traits, such as color and size. When Darwin's theory of evolution was published in 1859, it gave ornithologists a rationale for grouping species according to common ancestry. Birds were organized into families, and families into orders, based on presumed evolutionary history. But, because some taxonomists emphasize differences and others emphasize similarities, there is still no single agreed classification. Different classifications place birds in between 160 and 200 families, in 24 to 31 orders. Our knowledge of bird relationships, which affects the number of orders, families and species, changes as new information comes to light.

↑ **Swedish naturalist Carl Linnaeus** (1707–1778) devised the biological system of nomenclature. His naming system, *Systema Naturae*, was first published in 1735 and is the basis of all modern taxonomy.

↓ **To avoid confusion**, scientists assign a unique two-part Latinized name to every known bird. Closely related species are grouped in one genus; closely related genera are grouped in one family; and closely related families are grouped in one order. The resulting sequence represents the "tree" of evolution, tables diversity and provides the foundation for all biological knowledge.

KINGDOM **Animalia** Animal Kingdom *Eagle Owl, Blue Jay, lion, white shark, brown snake, stick insect, human, jellyfish, sand worm, sponge*

PHYLUM **Chordata** Animals with Notochord *Eagle Owl, kookaburra, lion, white shark, brown snake, human, sea squirts, lancelets*

SUBPHYLUM **Vertebrata** Animals with Backbones *Eagle Owl, pigeon, gazelle, gorilla, swordfish, turtle, human, salamander*

SUPRACLASS **Tetrapoda** Birds, Mammals, Reptiles and Amphibians *Eagle Owl, heron, gull, bear, pig, crocodile, frog*

CLASS **Aves** All birds *Eagle Owl, Ostrich, vulture, chicken, egret, kiwi, cassowary*

ORDER **Strigiformes** All owls *Eagle Owl, Barn Owl, grass owls, screech owls*

FAMILY **Strigidae** Typical owls *Eagle Owl, Great Horned Owl, Eurasian Pygmy-owl*

GENUS ***Bubo*** *Eagle Owl, Great Horned Owl*

SPECIES ***Bubo bubo*** *Eagle Owl*

EVOLUTIONARY PATTERNS

The science of classification is called taxonomy. The arrangement of the species in the class Aves into orders and families is not only a list of relationships, but reflects hypotheses about how birds are related. Phylogeny is the actual evolutionary relationships that taxonomists hope to represent in groups from a single ancestor. Ideally, all members of one bird family should be more closely related to each other than to any member of another family. New information, often biochemical in nature, prompts taxonomists to rethink the relationships among orders, and within orders, that indicate patterns of adaptive radiation. In a sense, taxonomists are reconstructing the evolutionary history of birds, and they use conservative characteristics or shared derived characters that provide clues to potential shared ancestors. However, taxonomists must always be aware of convergence, where a given trait is similar among different birds not because they share a common ancestor, but because they evolved the same solution to a common problem. A range of techniques has been used in classification, including anatomical traits, plumage patterns, behavior, analysis of egg white proteins, serum proteins, enzymes and DNA. The comparison of DNA (two strands of which are shown below) among species usually confirms the relationships among families based on morphology. Sometimes, however, new patterns emerge. These provide new hypotheses about bird relationships.

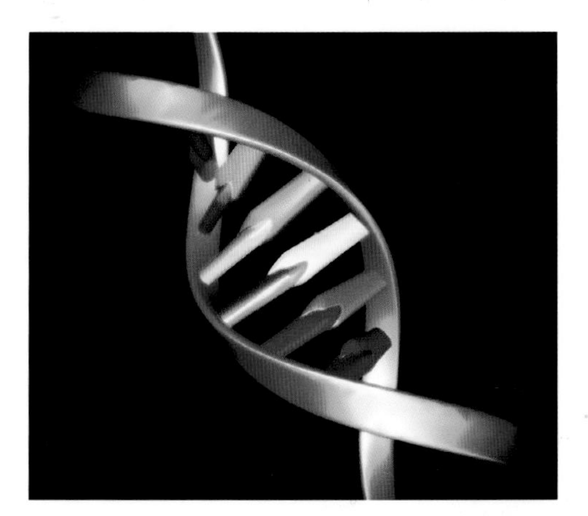

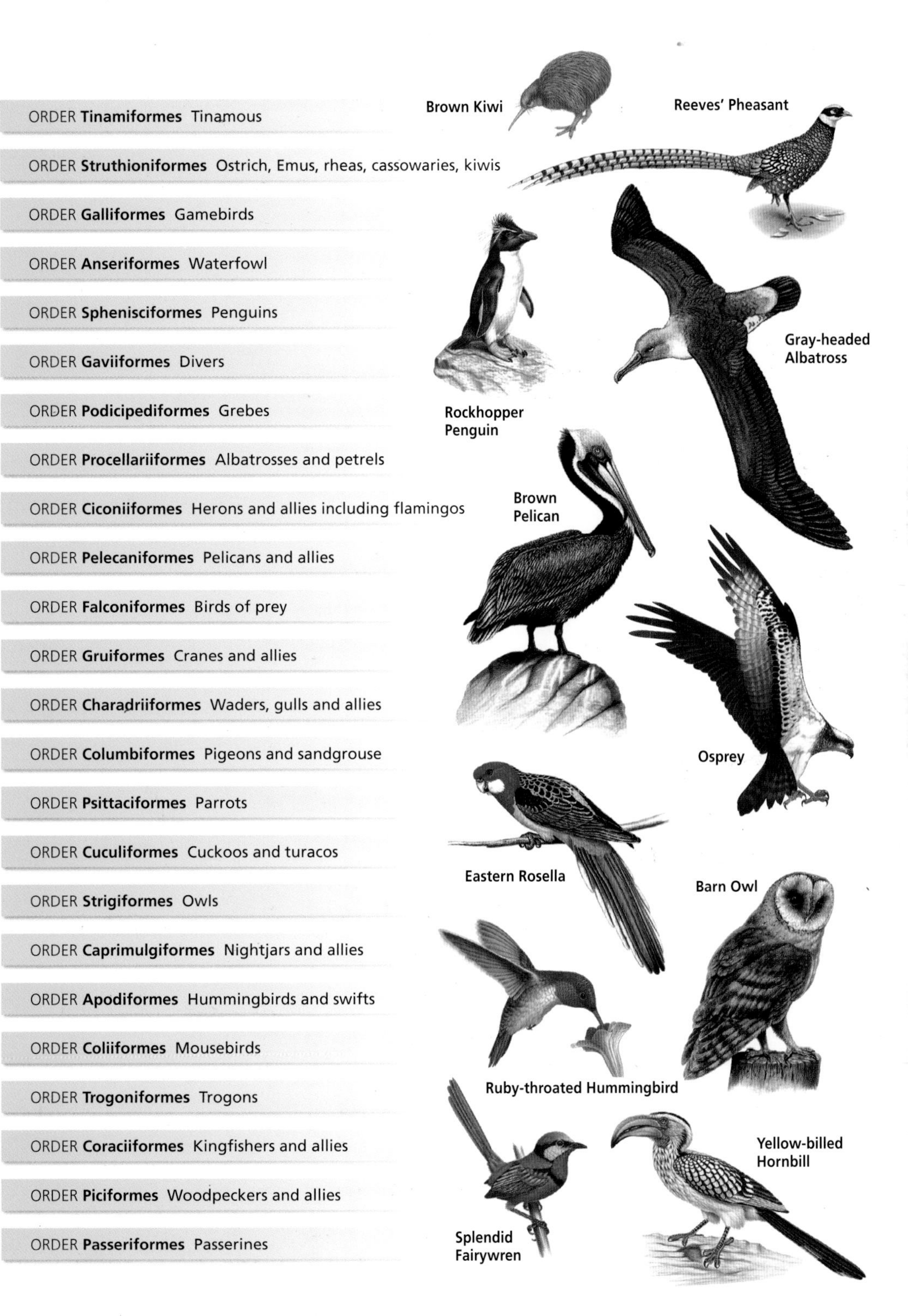

Ratites and tinamous

Class	Aves
Orders	2
Families	5
Genera	15
Species	60

Ratites are flightless birds that lack a keel on their sternum. They include Ostriches, rheas, Emus, cassowaries and kiwis, and are limited to the Southern Hemisphere. Ostriches are endemic to Africa; rheas to South America; Emus to Australia; cassowaries to Australia and New Guinea; and kiwis to New Zealand. Most ratites are large, long-legged, diurnal birds. Kiwis, the exception, are nocturnal forest birds with relatively short legs. Ostriches, rheas and Emus live in grasslands and savannas while cassowaries inhabit forests. Ratites' bodies are covered in shaggy feathers, but (except for kiwis) their necks and heads are almost bare. Males usually incubate the eggs and provide most or all the parental care, while females seek other mates. Young are precocial and leave the nest soon after hatching. They remain with parents until they are able to find food and shelter, and avoid predators by themselves.

Although they are fully capable of flight, tinamous usually run from danger. They are placed in their own order and are found from Southern Mexico, through Central and South America to Patagonia. They are protectively colored to blend in with the forest, and like the ratites, males do most of the incubating.

→ **Male Emus**, which stand nearly 6 feet (2 m) tall, stay with their precocial chicks to lead them to water and food, and to protect them from predators. As well as running quickly, Emus are good swimmers.

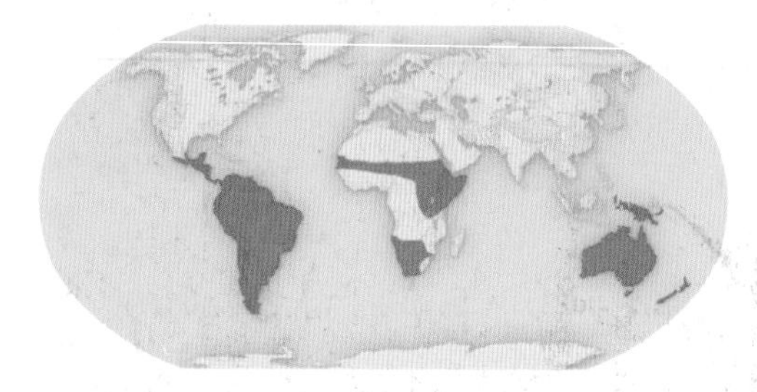

AVOIDING PREDATORS

Flightless birds have evolved special strategies to avoid predators. Kiwis have solved the predator problem by being cryptic and nocturnal. Other ratites are large, swift and strong, able to outrun predators or attack them physically. When chicks are threatened, however, their first defense is to crouch motionless in the grass. Chicks are cryptic, with streaked and mottled plumage, and remain hidden when given the signal by their parents. Ostriches, rheas and Emus may protect their eggs and chicks with elaborate distraction displays, drawing predators away from their offspring. Cassowaries rely on outrunning their predators in thick forest vegetation, and the casque on the top of their heads protects them from branches. Kiwis lay their eggs in underground burrows for protection. Tinamous rely on being cryptic in dark forests; their chicks are also cryptically colored to blend in with the forest floor, but surprisingly, their eggs are not cryptic at all.

← **Kiwis are the only birds whose nostrils** open at the tip of the bill, giving them a keen sense of smell to find grubs and worms. Their short legs are placed so far apart that they waddle when they run, and they have neither wing nor tail plumes. This Great Spotted Kiwi is feeding on a worm.

↓ **Ostriches evade predators** on the African savanna by outrunning them. They also travel with ungulate herds as an additional anti-predator strategy. Here, a male with black plumage is running with two females, showing off their powerful legs.

Gamebirds

Class	Aves
Orders	1
Families	5
Genera	80
Species	290

Gamebirds are a large group of chicken-like species that include pheasants, grouse, quail, turkeys, curassows, guans, chickens, chachalacas, guineafowl and the curious megapodes. Most gamebirds have been hunted relentlessly, and many are endangered. However, several species, such as chickens, turkeys and quail, have been domesticated. About two-thirds of all gamebirds are pheasants—short-winged ground birds—which have diversified into spectacular long-tailed species in Asia. Many gamebirds make loud, far-reaching and raucous sounds, especially in the morning. For example, wood quail of tropical America greet the dawn with loud, rollicking duets as they go to roost. Megapodes, such as incubator birds and brush turkeys, live in Australia, New Guinea, Indonesia, the Philippines and other South Pacific islands. They have the unique habit of mound nesting. Using their large feet, megapodes scrape vegetation into mounds that can be up to 10 feet (3 m) tall and 25 feet (8 m) wide. They deposit their eggs in the mound, where heat from the rotting vegetation incubates them. Parents stick their heads in periodically to monitor nest temperature, adding or removing vegetation as required. The young molt from down to feathers while they are still inside the egg. They dig their way out when they hatch, and fly immediately away, often without ever seeing their parents.

This white mutation of the Indian Peafowl is one of many that have been developed in captivity. In this mutation, the genetic mechanism that creates pigments in the feather barbules has been shut off, so that impinging light is reflected as "white" light. Only males have the spectacular tail fan seen here in display; the females are drab.

Two Vulturine Guineafowl in Samburu National Reserve, Kenya, raise their heads to watch for danger. In contrast to most other gamebirds, male and female guineafowl look so similar that it is difficult to tell them apart.

Helmeted Guineafowl travel to waterholes in the Okavango Delta of Botswana in large groups of as many as 100 birds. Some flock members stand guard and keep watch for predators while others stop to drink.

↑ **A Capercaillie male in flight** is particularly attractive to female birds that lurk in the shadows below.

↑ **Most female pheasants** have dull, cryptic plumage that blends in with the fields and forest floors on which they nest. A harvester has recently passed in this field, revealing a female nestled in the vegetation.

ELABORATE COURTSHIP

Males of many gamebird species stage elaborate courtship displays. They use their brightly colored plumage, topknots, ruffs, throat sacs, elongated tail feathers and tails, which they spread and display to attract females. Gamebirds also call for mates. Guans call from the trees; grouse beat their wings to drum from logs; prairie-chickens inflate their orange air sacs and boom; Sharp-tailed Grouse stomp on their small leks; and Wild Turkeys fan their tails, drop their wings and give low, thundering calls known as gobbles. Male pheasants, especially in Asia, are particularly attractive. Golden and Silver pheasants have breathtakingly beautiful plumage. The Golden Pheasant has a bright red belly, a golden crown and back, and a long golden tail; the Silver Pheasant has a red face, elongated black crown feathers, spectacular white feathers edged with black on the back, and elongated, black-barred white tail feathers. The long or ornamented tails of species such as the Great Argus Pheasant and Common Peafowl are rivaled only by the hummingbirds.

Waterfowl

Class	Aves
Orders	1
Families	2–3
Genera	52
Species	162

Waterfowl live in lakes, ponds and rivers, and along coastal marshes and bays. Geese, swans, ducks and screamers make up this order, and most are sought after as game birds for food and feathers. Except for screamers, all waterfowl have broad bills and most are strong flyers. Waterfowl chicks leave the nest after hatching and, although they can feed on their own, remain with their parents for protection until fully grown. Diet varies greatly, from plants and invertebrates to small fish. Geese and swans mate for life and even migrate together once they have paired. Both parents incubate and care for their young. Ducks, in contrast, are polygamous and separate immediately after copulation. Most ducks build nests on the ground or over water but there are notable exceptions: Wood Ducks nest in tree cavities and Black-headed Ducks deposit their eggs in the nests of other ducks, eliminating all parental duties. Screamers are large, gooselike gray birds that live only in South America. Like penguins, they have feathers over all their bodies. They are birds of marshes, wet grasslands and forest lakes, and usually nest in large wetlands, away from ground predators. Screamers are named for their trumpeting call, given day or night, which carries far over the pampas.

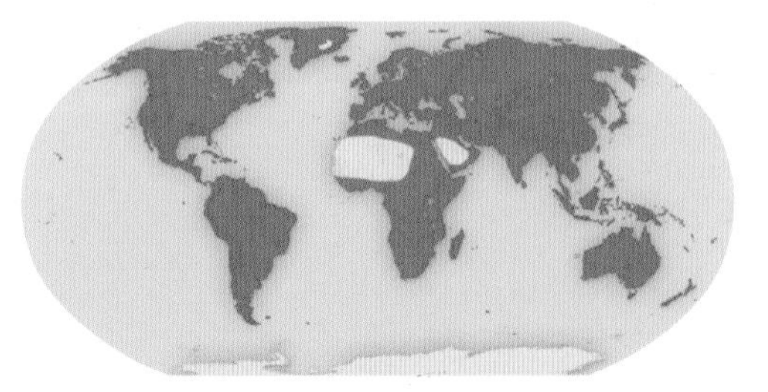

→ **During courtship displays**, Mandarin Ducks dip their bills in the water, erect their head feathers or crests and raise their tails or back feathers.

↓ **Mallards and other ducks** form dense resting or roosting flocks on lakes and reservoirs. When they take off it is difficult for predators to single out individuals.

↑ **Most geese fly in V-formation.** The leader breaks the air and causes a streamlining effect, stirring up updrafts that reduce air resistance for the birds flying behind. Each bird takes its turn flying in the lead so that they may all conserve equal amounts of energy.

← **Mute Swans are attentive** to chicks, brooding them under their wings to keep them warm and safe from predators. Parents protect chicks aggressively. They bite intruders and flap their wings wildly.

↙ **All waterfowl have webbed feet**, which help them to swim on the surface as well as underwater.

DUCK DISPLAYS

Courtship in waterfowl varies, depending on the pair bond. A waterfowl's plumage reflects its mating system. Swans, geese and screamers mate for life, and males and females look the same. Most male ducks, however, are brightly colored during the mating season, as they continually display and court different females. Duck displays are ritualized and many evolved from behaviors originally meant for preening, drinking or bathing. Many ducks elevate head or crest feathers, or back or tail feathers during displays. In most species it is the male that works the hardest, calling and displaying to females until one appears interested. Once a female is interested, they engage in mutual displays involving dipping their bills or heads in the water, pulling their bills into their chests, aborted preening motions or rising slowly from the water. Ruddy Ducks have some of the most elaborate displays. They erect and spread their tails, which is why they are commonly known as "stiff tails." They then lower their heads and steam rapidly across the water in front of an alluring female. At other times, they inflate their chests, pull their heads in toward their chests, and blow bubbles into the water.

Penguins

Class	Aves
Orders	1
Families	1
Genera	6
Species	17

Penguins are flightless seabirds confined to the Southern Hemisphere. They nest on the ground in large, dense colonies on islands and mainland coasts. Several species live on the ice floes of Antarctica, and nest on the continent's bleak, windswept landscape. Penguins' wings are modified into flippers. This allows them to "fly" through the water in pursuit of prey. They have a large keel to which swimming muscles are attached, and their webbed feet are used for steering underwater. Because their legs are so far back on their bodies, penguins walk upright, endearing them to people everywhere. Since they do not fly, there is no advantage in having air sacs in their bones, which are solid. Unlike most other birds, which grow their feathers in tracts, the plumage of penguins is continuous and the feathers are small to provide extra insulation. A penguin's plumage is usually white on the belly to prevent prey seeing it from below, and gray or black on the head and back to prevent predators seeing it from above. All penguins are highly gregarious and travel and forage in groups at sea. They feed on fish, shrimp, squid and krill. Both parents usually incubate eggs. Young stay in the nest for many weeks or months and are fed by their parents.

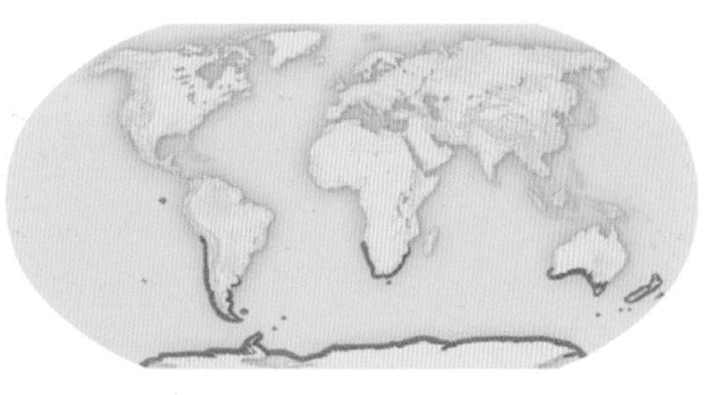

↓ **King Penguins are streamlined** for swimming rapidly in pursuit of prey. They use their small, flipper-like wings to propel themselves underwater, and their feet to steer. When chasing fish and squid, King Penguins can dive to depths of more than 1000 feet (300 m).

EXTREME NESTING

Penguins live in some of the harshest environments on Earth. Emperor Penguins never set foot on bare land; they spend their lives in cold Antarctic waters in the broken icepack, feeding on krill. They start breeding in the autumn, when pairs return to their old nest sites. In late autumn, just as winter approaches, the female lays her single egg, then heads out to sea to forage, leaving the male to incubate alone for more than two months. Males huddle together, moving in and out of groups to stay warm during the winter, when temperatures drop to -40 degrees. They lose about a third of their weight by the time their chicks hatch, but retain a small meal to regurgitate to the hatchling. Within a day or two of hatching, the fat, sleek females toboggan in to feed the chicks and relieve the males, who head out to sea. Males take a few weeks to regain their weight and then return to help the females feed the young until fledging.

↑ **The world's largest King Penguin colony** is on Macquarie Island in the subantarctic. Once almost eliminated by the oil trade, penguins are now protected here and number more than half a million.

→ **An Emperor Penguin parent** looks for its chick among a crèche of young. Adults incubate through the winter so chicks have the maximum benefit of a long summer feeding period before independence.

↓ **A penguin's wings** are covered by a dense layer of short, stiff feathers. Fluffy down at the feather base traps warm air, while oily feather tips keep seawater out. The wing bones are short, thick and flat, forming a flipper-like surface for propulsion underwater.

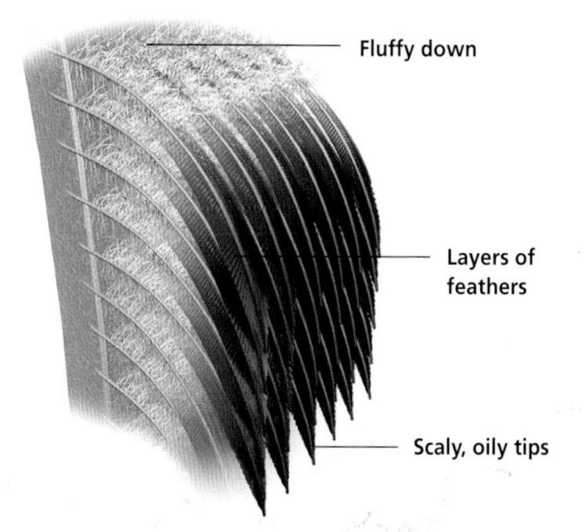

Divers and grebes

Class	Aves
Orders	2
Families	2
Genera	7
Species	27

The plaintive, mournful call of divers, also known as loons, resounds over lakes in the northern parts of Eurasia and North America where they nest. Because their legs are placed far back on their body, although not as far back as penguins, they walk awkwardly on land. Divers have fully webbed feet, used to propel them underwater. They go ashore only to nest, because they can barely walk, and place their nests close to the water's edge. Both parents incubate and care for the young, which leave the nest soon after hatching and, like grebes, ride on their parents' backs for protection. They eat mainly fish, as well as mollusks, crustaceans and other aquatic invertebrates.

Grebes are strong and agile swimmers, but weak flyers. They have short, stubby wings and tails. Like divers, they are awkward on land and propel themselves underwater with their feet. Grebes inhabit inland lakes and ponds. They are monogamous and parents take turns incubating. Their nests are built over water, a strategy that places them away from most ground predators. When a predator does approach, the incubating adult pulls nest material over the eggs and slips quietly into the water, submerging, then resurfacing away from the nest. The nest is a wet, decaying mass that must be continually added to, to prevent it sinking below the water's surface. Grebes eat their own feathers and feed them to their young. This may serve to line the stomach and prevent punctures from sharp fish bones. Apart from fish, they also feed on aquatic invertebrates and some vegetation.

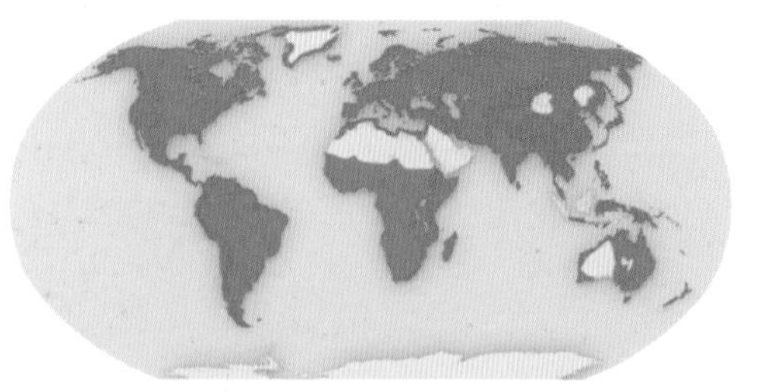

→ **During an aggressive encounter**, or when they see an approaching intruder, Common Loons rise up in the water to defend their young.

↓ **Red-necked Grebes** build massive nests of wet and decaying vegetation which they must continually replenish, or the nest will sink. Nests are usually attached to vegetation to prevent them from floating away during strong winds, especially in boreal Alaska.

GREBE DISPLAYS AND DIVER CALLS

Grebes perform spectacular courtship displays on the water. Western Grebe pairs swim side by side, and as they swim faster, their bodies become more upright until they run in tandem across the water. In the "weed dance," Great Crested Grebes float, breast to breast, with their feathers erect and heads wagging. They rise out of the water and face one another, holding vegetation in their beaks, then sink slowly. Most other grebes raise their brightly colored breeding plumes, known as horns, ears or collars, when they engage in their water dances.

All divers (loons) are extremely vocal, especially during courtship and at night. They give a series of loud calls, yodels, tremolos and wails. Long, drawn out calls rise in the middle and fall toward the end, like a wolf's howl. The calls can sound like eerie, maniacal laughter, and gave rise to the phrase "crazy as a loon."

↑ **Red-throated Divers** breed on tundra lakes. They have solid, heavy bones, which reduce buoyancy and allow them to sink slowly into the water. These birds are noted for their diving ability in search of fish, their primary food.

→ **Young grebe chicks** ride around on the backs of their parents for protection. Here, a Horned Grebe chick sits on its parent's back while the incubating parent waits for the other eggs in the clutch to hatch progressively. Its mate may have just fed the young chick some vegetation.

Albatrosses and petrels

Class	Aves
Orders	1
Families	4
Genera	26
Species	112

Albatrosses, petrels, shearwaters and fulmars are tube-nosed birds that soar over the open oceans. They spend more time over the trackless seas than any other birds. When albatrosses fledge from their nests, they leave and may not return to land for six or seven years. They fly continuously over the open ocean, occasionally landing to rest. Birds in this group are able to drink seawater, and remove the sea salt with their salt gland, which is located at the base of the bill in a depression on the skull. The salt drips out of their bills through their tubed noses. These birds range widely in size, from great albatrosses to smaller storm-petrels and diving petrels. As an order, they exhibit many feeding methods. Giant Fulmars are significant predators on penguins. Albatrosses pick up fish and squid from the surface. They may land near feeding pods of seals, hoping to find discarded bits of fish, or trail behind boats to snatch fish churned to the surface. Smaller storm-petrels skim along the water, pattering with their feet to grab items from the surface. Diving petrels dive far below the surface. Tube-noses mate for life, which in the case of albatrosses, can be 50 or 60 years.

→ **All albatrosses, such as this Black-browed Albatross**, spend most of their time soaring over the ocean on updrafts in search of food. They walk awkwardly, and come to land only to breed.

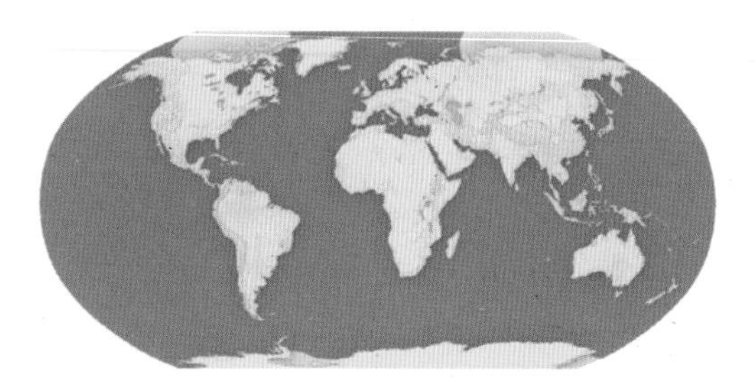

⇠ **Black-browed Albatrosses** nest in large, dense colonies, such as this one in the Falkland Islands. Chicks remain in the nest for weeks or months, waiting for parents to bring back food. Some albatrosses fly hundreds of miles in search of fish and squid for chicks.

← **Laysan Albatrosses** engage in a ritualized courtship display where the pair dip and bow to one another, and then cross their heads in a deep swoon.

→ **This map shows** the paths of Wandering Albatrosses as they ranged across the ocean from nesting sites on Crozet Island. The birds carried small transmitters, which allowed researchers to track them by satellite.

↓ **Black-footed Albatrosses** court one another with elaborate dances, nasal honks and crooning calls. An observer crouches low to appease the displaying pair.

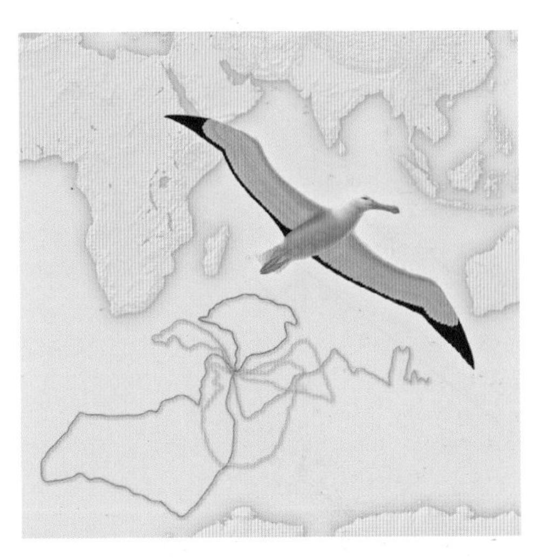

PROLONGED COURTSHIP

Long-lived, monogamous birds must pick their mates carefully, and some tube-noses devote years to finding just the right mate. Because of their difficult foraging method, breeding is delayed until the birds can hunt efficiently enough to feed young. Most albatrosses initiate breeding between four and ten years, while petrels begin breeding at four to six years. The Campbell Royal Albatross delays breeding until it is ten years old, but comes to land earlier to search for a mate and find a territory. All tube-noses breed on islands, removed from mammalian predators, in small to large colonies. Courtship is an elaborate and prolonged affair, particularly for albatrosses. Male albatrosses arrive early, select a site on a grassy hillside and begin to display to the wind. They give loud trumpeting calls, point their bills skyward and extend their wings, trying to entice a female to land. This can go on for days, weeks, and even years. Sometimes four or five birds engage in ritualized displays as males compete for females. Dancing groups can erupt into overt attacks if one pair feels their territory has been invaded.

Flamingos

Class	Aves
Orders	1
Families	1
Genera	3
Species	5

Brilliantly pink, with long legs and necks, flamingos are one of the world's most picturesque birds. Taxonomists never know exactly where to place these unique birds on the phylogenetic tree. Sometimes the five species are placed with the herons, but flamingos also share an ancient bird ancestor with waterfowl. They breed in shallow lakes and lagoons of tropical West Indies, South America, Asia, Europe and Africa, and often prefer brackish water. Monogamous birds, they build mound nests of mud, usually on the edge of lakes or in shallow water. The clutch of one to two eggs is incubated by both parents, who then continue to feed the chicks until they are able to fly. Flamingos eat a wide range of food, such as algae, diatoms, protozoans, small worms, aquatic plants, insect larvae and even small crustaceans, which they strain from the water. The ancient Romans considered flamingo tongues a great delicacy.

NESTS AND CRÈCHES

Flamingos build conelike mud nests to keep their eggs and young chicks above the shallow waters of inland lakes where they nest. The cones can be up to 20 inches (50 cm) high, and twice as wide at the base. The soft mud hardens and supports the flamingo's weight during incubation. The nests are placed close together, since they are used mainly during incubation and there is little aggression between neighbors. Within a few days of hatching, the chicks are strong enough to leave the nest and form large groups called crèches. If predators approach, they run quickly to the water and swim away, still in their crèches. By three weeks, they forage alone and are almost independent.

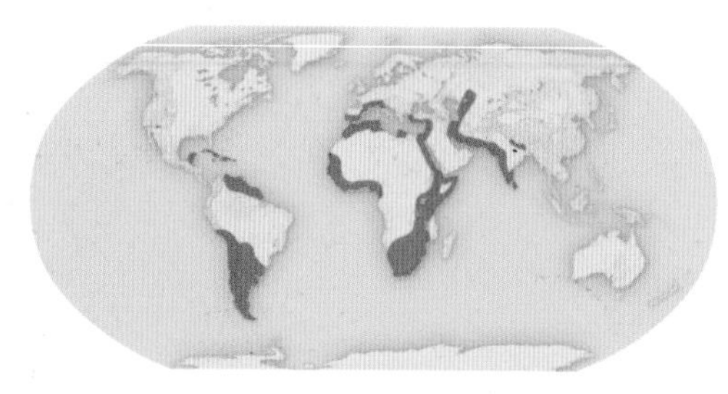

↑ **A flamingo feeds** by straining tiny organisms from shallow water and mud. Pumping actions of the throat and rapid movements of the tongue suck water through the mouth and over small, hairlike plates around the bill, which strain the food.

← **A flamingo's lower bill** acts like a trough. Its upper bill acts like a lid and lies almost flat on the mud while feeding.

↓ **During courtship**, Greater Flamingos fly in tight groups. They honk much like geese.

Herons and allies

Class	Aves
Orders	1
Families	3–5
Genera	41
Species	118

Herons and their allies are generally large, long-legged waders with long, heavy bills. This diverse group includes herons, egrets, ibises, bitterns and storks. Based on biochemical evidence, some taxonomists even include the New World vultures in this order. Although waders are traditionally associated with water, some herons and ibises have adapted their feeding to dry fields. Most herons and egrets feed mainly on fish, stabbing into the water with their long necks and long, strong bills. Storks, such as Hamerkops, have a similar build but with heavier bodies and longer bills. Many feed mainly in terrestrial habitats on herptiles and small mammals. Marabous have unfeathered heads, an adaptation to feeding on carrion. Ibises have a curved bill and spoonbills have a flattened, spatulate-like bill. The most unusual member of the group is the Shoebill of Africa, also known as the Whale-headed Stork. It has a huge head and a bill shaped like a wooden shoe. Herons and their allies are all monogamous. Most nest in colonies and both parents incubate and care for the young. Chicks are helpless when born and require continual attention and protection during the brooding period. Most species return to the same breeding sites year after year, and colonies may be occupied for centuries.

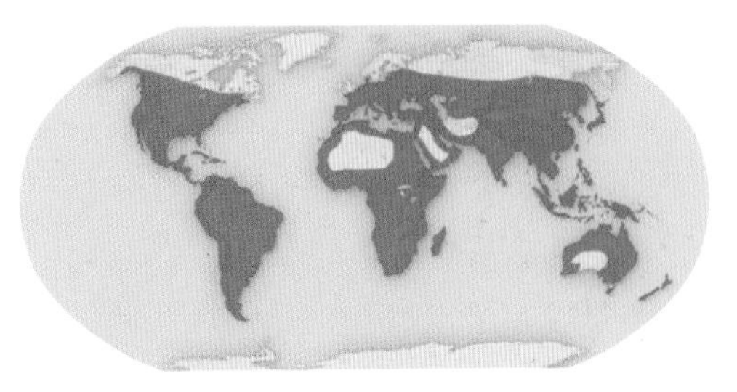

↓ **Great Blue Herons** use their long necks and bills to stab at fish in shallow water. They are stand-and-wait predators, remaining motionless for hours while they wait for an unsuspecting fish to swim by. With a quick thrust they seize the fish, then flip it with a shake of their bill before swallowing the fish headfirst to avoid being injured by its sharp scales.

Herons and allies continued

NESTING HIERARCHY

Some birds like company, especially while nesting. Herons, egrets, ibises and storks often nest in mixed-species colonies with each other, and with other species such as cormorants and gulls. In mixed-species colonies, there is an order to nesting patterns. Larger species generally nest higher in the shrubs or trees than smaller species, giving them the best vantage point. This nesting strategy has been established over a long, shared history and reduces the overall competition for space. Large herons, such as the Gray Heron in Europe and Africa and the Great Blue Heron in North America, nest on the highest choice spots, far removed from predators. Small birds, such as Little Egrets in Europe and Snowy Egrets in North America, nest in the lowest foliage. Cattle Egrets are an interesting exception. In their native home of southern and central Africa, they nest in the height order that their size would predict. But over time, Cattle Egrets have spread to the rest of Africa, Europe, Asia and North and South America. During their expansion, they became more aggressive than native species, and in these new regions are able to place their nests higher in trees than their size would predict.

↖ **Male Great Blue Herons** bring branches to their nest sites as courtship offerings. Once the pair has a clutch, the male continues to bring offerings to cement the bond.

↑ **Great Egrets lay their eggs** on successive days, but begin incubating after the first egg, so chicks hatch progressively. Older chicks can beg more vigorously and lift their bills higher than younger chicks to obtain more than their share of food.

↗ **Roseate Spoonbills** walk slowly with their half-open bills in water. They swing their bills from side to side, closing them on anything edible.

→ **Great Blue Herons** usually nest in colonies located in tall trees, or on inaccessible islands, to prevent predation by mammals. They build their bulky stick nests in the crotches of trees for stability, but strong winds or storms may dislodge them.

← **Jabiru Storks** wade through shallow water in the Pantanal of Brazil. They peer into the water as they stalk the wetland shallows, waiting to spear fish.

Pelicans and allies

Class	Aves
Orders	1
Families	6
Genera	8
Species	63

Pelicans and their allies include such diverse birds as pelicans, tropicbirds, boobies, cormorants, anhingas and frigatebirds. Pelicans are the most familiar species, because of their large throat sac. Birds in this group breed in dense colonies on inland lakes, along coasts and on oceanic islands, and often nest with other species. Many are monogamous and mate for life, devoting considerable time to courtship and finding a mate. Like most other monogamous species, male and female birds generally look alike. There are exceptions: female anhingas have a golden head and neck while males are all black; and male frigatebirds have a red throat sac which they inflate during courtship. Males select a nest site and display to prospective females, but both sexes actively engage in courtship. They call, scrape, bow, touch bills and dance about the territory. These birds nest in a range of habitats, on the ground, in low shrubs or in trees, and usually build nests from sticks or mud. Clutch size ranges from one to six eggs, depending upon the species. Both parents incubate and care for the chicks. In some species of frigatebirds and boobies, it can be as long as 170 days from the time chicks hatch until they leave the nest.

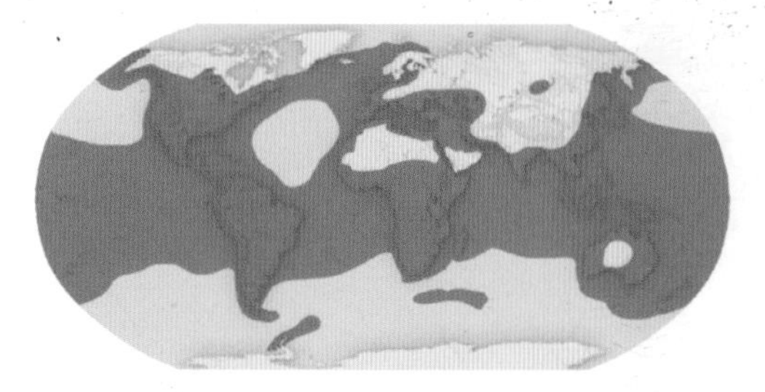

↓ **Australian Pelicans** sometimes form dense feeding frenzies where fish concentrate or where fishermen have tossed offal to them. Gulls join the foray, hoping to steal food from the pelicans. Here, the expanded throat sac can be clearly seen on the center pelican.

↑ **Great Cormorant** young plunge their bills deep into the mouths of their parents to pull out food that the parents regurgitate. Cormorants are also known as shags.

↗ **African Darters** swim partly submerged with only their heads above water, like a submarine. They dive from the surface to spear fish, such as this bream.

→ **Red-footed Boobies** sometimes build stick nests in trees on tropical atolls to remove chicks, such as this one, from disturbances by other colony members and from ground predators. When nesting on the ground, a booby's nest is not much more than a circle of excreta.

FORAGING FOR FISH

Pelicans and their allies have a wide range of feeding methods, but they mainly eat fish. Some birds plunge-dive from great heights; some pick items off the water surface; others dive from the surface to spear fish underwater. Brown Pelicans fish by plunge-diving. White Pelicans swim in great flocks, corralling and driving fish schools into shallow water and scooping them up with their great bills. Sometimes pelicans feed cooperatively by forming dense V-shaped platoons of hundreds of birds that skim over the water toward shore, forcing the fish into shallow water with their beating feet and wings and making them easier to catch. A pelican's bill can hold three times as much as its stomach. Tropicbirds and boobies feed mainly on fish and squid, which they obtain by plunge-diving. Anhingas and cormorants, the least marine of the group, swim in the water and dive to spear fish. Underwater, they propel themselves mainly with their feet, assisted by their half-opened wings. They can dive to great depths. Frigatebirds are the pirates of the sea, pursuing other seabirds to steal fish from them. Adult frigatebirds are more adept than young at piracy, also called kleptoparasitism. They chase young gulls, terns, boobies, cormorants and pelicans, forcing them to drop or disgorge their food, and then catch the food in midair before it hits the water.

Birds of prey

Class	Aves
Orders	1
Families	4
Genera	83
Species	304

Diurnal birds of prey include hawks, eagles, vultures, osprey, Secretary-birds, falcons and caracaras. With their sharply hooked bills, strong bodies, powerful feet and sharp claws, birds of prey have a familiar form that is recognized the world over. All are monogamous, and keep the same territory and nest site from year to year. Males are significantly smaller than females in most species. This is reverse sexual dimorphism—in most dimorphic birds the male is larger. Although both sexes incubate, it is mainly the female that guards the young, leaving the male to hunt and bring back food for the chicks. Ornithologists speculate that females are larger so they are able to defend their young against the male, whose hunting and killing instincts might be stimulated by seeing its own offspring. Birds of prey have figured prominently in folklore, legend and religion, and have been hunted, trained for falconry and revered throughout the ages. They are strong fliers that feed mainly on animal matter and carrion. This has made them particularly vulnerable to pesticides, heavy metals and other contaminants because they eat animals that are fairly high on the food chain, and have themselves accumulated contaminants. Many species declined when DDT was in use.

→ **White-backed Vultures** sit on top of a dead elephant in Chobe National Park, Botswana. The skin is too tough for them to break, so they must wait for other vultures, hyenas or lions to come and open the carcass.

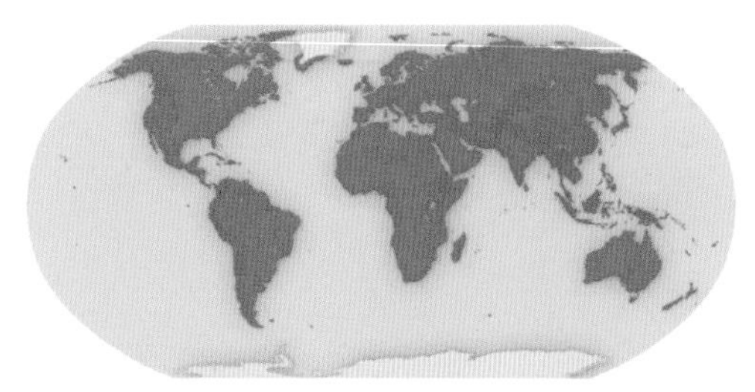

LIVE PREY

Most birds of prey feed on live animals, using their strong, hooked bills to tear flesh. Hawks and eagles eat mammals, reptiles, fish and other birds that they mainly catch on the ground. Their close relatives, the Old World vultures, eat carrion. Some eagles, such as Bald Eagles, feed on salmon and other fish that die after spawning. Falcons are noted for their fast dives in pursuit of flying terns, shorebirds and other small birds. Peregrines have been clocked diving faster than 200 miles per hour (320 km/h). Ospreys feed only on fish, diving to grasp the fish in their strong talons. They then carry the fish to a safe place before tearing into it. Secretary-birds are unique in that they walk on the ground in search of snakes, lizards and young birds. New World vultures, or condors—a group whose taxonomic relationships are unclear—are large soaring birds that eat carrion.

→ **The African Harrier-hawk**, or Gymnogene, is an accomplished acrobat. It clings to branches or nests with one leg and uses its other leg to extract young birds from nests. Its extendible legs give it an extraordinary range of motion.

↟ **In winter, when food is hard to come by**, birds such as these Common Buzzards must fight over carrion in the snow.

↑ **Montagus Harriers** engage in elaborate aerial courtship rituals which involve circling one another. Here, a male flies above a female. As part of their courtship, the male passes a prey item to the female.

← **The Snail Kite's slender bill** is adapted to cut a snail's retractor muscle and extract it from its shell. The apple snail is its main prey. Snail Kites favor deserted perches where other birds will not try to steal their prey, and a pile of empty snail shells often accumulates under these perches.

Birds of prey continued

LENGTH OF WINGSPAN

Andean Condor
11.5 ft (3.5 m)

Bald Eagle
7.5 ft (2.25 m)

Turkey Vulture
6 ft (2 m)

Osprey
5–6 ft (1.5–2 m)

Black Vulture
4.5–5 ft (1.25–1.5 m)

Gyrfalcon
4–4.5 ft (1–1.25 m)

Peregrine Falcon
2.25 ft (0.7 m)

American Kestrel
2 ft (0.6 m)

↑ **The shapes and sizes** of birds of prey are diagnostic, allowing bird-watchers to quickly place them into families. Their wings reflect different modes of flight and capturing prey. Large, broad wings soar on updrafts; thin, long wings allow rapid diving in pursuit of fast-moving prey; and smaller, rounded wings permit birds to move through thick vegetation. Different wingspans are listed above.

SOLITARY NESTING

Most birds of prey are loners; breeding pairs need large territories in which they hunt alone for prey. They generally build stick nests in trees, on cliffs or on the ground, and use the same nest each year—simply adding a few more twigs and fixing up the lining with smaller twigs or grasses. Sites are carefully selected, out of the reach of mammalian predators. The territory must include a suitable nest site, a perch site and sufficient space to hunt for food. Many species have adapted to nesting in civilization.

↑ **A Bald Eagle** clutches a fish with its talons while ripping it apart. Eagles feed mainly on fish, which they pluck out of the water in smooth dives.

→ **An adult Steller's Sea Eagle** flies to its nest and waiting eaglet in Bolshoy Shantar, Russia. The strong nest platform is almost as large as a king-size bed.

← **In many areas,** such as coastal islands or waterways, Ospreys nest on artificial structures, such as this pole.

Cranes and allies

Class	Aves
Orders	1 or 2
Families	12
Genera	62
Species	213

Cranes, rails, bustards, trumpeters, finfoots, Limpkin and Kagu make up this diverse group. Cranes are large, long-necked, long-legged, gregarious and loud. They mate for life and have long lives; in Japan the crane is a symbol of longevity. In contrast, rails are small, short-necked, short-legged, secretive and quiet. They rely on cryptic coloration to protect them against predators, and the smaller rails live for only a few years. All cranes and their allies are omnivorous, although their bill shapes differ. Cranes have intermediate to long, sturdy bills; rails have shorter, curved, probing bills; and the bills in the other species reflect their more specialized feeding habits. Limpkins have long, slightly curved bills for probing the mud for mollusks, crayfish and worms; finfoots use their short bills to catch fish. Virtually all members of this group are, or have been, exploited for food and many are now endangered because of hunting pressure and loss of habitat. Several species of flightless rails that evolved on predator-free islands became extinct when humans introduced predators.

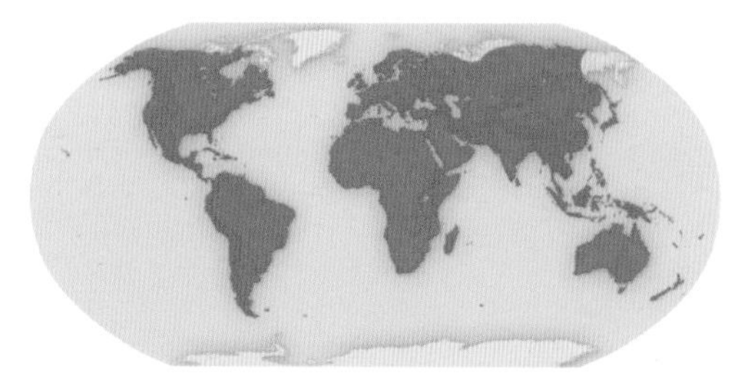

↓ **The elaborate courtship dances** of Japanese Cranes involve jumping up and down with bills pointed skyward and wings extended. Their giant courtship leaps take them far across snow-covered fields.

COURTSHIP DANCES

Courtship varies markedly within this group. The large, stately cranes engage in elaborate displays and dances of two or three individuals, or sometimes dozens, calling loudly. Many crane species dance for hours on end. They jump up and down, face one another, flap their wings wildly and call continually. Sandhill Cranes court during their migration in the prairies of central North America, pausing at traditional stopover areas, such as the Platte River in Nebraska, to feed in fields and roost in shallow waters. Trumpeters also travel through their South American forests in sociable flocks and engage in noisy, cranelike dances. In contrast, rails are solitary and usually secretive in their courtship.

← **Most rails, like this Water Rail**, wade slowly among the reeds. Their drab colors blend with the reeds and shadows. When they reach open water, rails dash to the next clump of vegetation and quickly disappear.

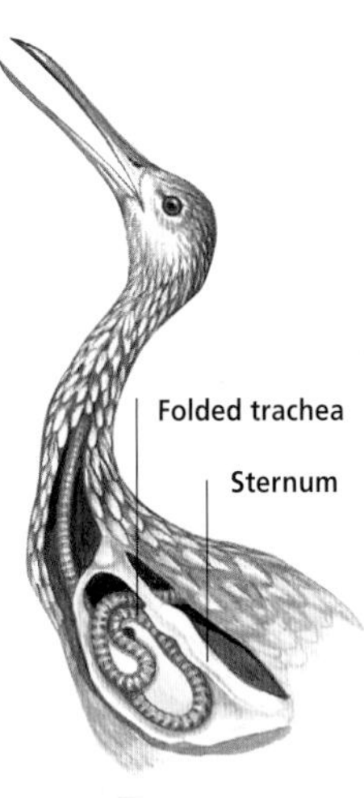

↑ **A crane's trachea** is coiled in its body, which allows it to give loud, resonant calls that can be heard several miles away. Male cranes call to attract other cranes to their courtship grounds.

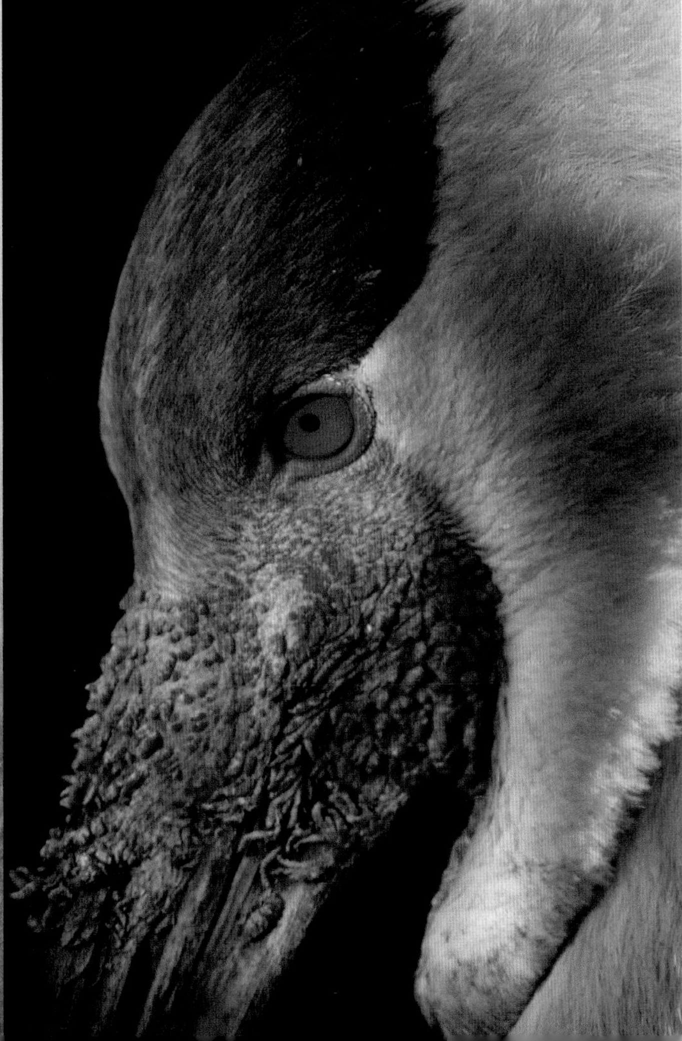

← **Wattled Cranes** are endemic to Africa. Only adults have red, bare skin under the eyes that extends to the tip of the wattles. This species is highly endangered.

↞ **Kori Bustards** tuck back their heads and fluff out their neck feathers to display to others birds or to repel intruders.

Waders, gulls and allies

Class	Aves
Orders	1
Families	16
Genera	86
Species	351

Shorebirds include oystercatchers, jacanas, sandpipers, plovers, avocets, stilts, phalaropes, gulls, terns, skimmers, auks and skuas. Most species live along beaches, estuaries, lakeshores and river banks, feeding mainly on fish and aquatic invertebrates. They have webbed or semiwebbed toes that aid swimming in water and walking on mud. Sandpipers and plovers are generally small birds that feed at the water's edge; terns mainly forage by plunge-diving for fish; and gulls are omnivorous, feeding on nearly every food type. Auks, such as puffins, are the Northern Hemisphere equivalent of penguins. They feed in the open ocean, diving for fish or crustacea and using their wings to propel themselves underwater. Unlike penguins, auks are fairly strong fliers. They generally nest in burrows in the ground or on cliff ledges. The young of sandpipers, plovers, avocets, stilts and phalaropes are highly precocial. Chicks leave the nest within hours of hatching and are able to find their own food. In contrast, the young of terns, skimmers and auks remain in or near the nest for weeks. They are fed and guarded by their parents, who bring food back to their chicks in their bills, or in the case of gulls, regurgitate to feed their young.

↓ **Great Black-backed Gulls** (left) often find it easier to fight over fish, shellfish or other prey that another bird has obtained (right), rather than finding their own. These gulls are predators and will kill adult ducks on the water or knock migrating songbirds out of the air, then pluck them from the sea surface.

→ **Sandpipers**, such as this Western Sandpiper, feed on mudflats, at the water's edge or in shallow water. Sometimes they foot-paddle to scare up invertebrates.

↓ **Razor-billed Auks** dive for fish and bring them back to the nest in their bills. They nest on the same narrow cliff ledges year after year. Both sexes incubate the eggs and feed the chicks.

Waders, gulls and allies continued

↑ **Long-tailed Skuas** nest in the tundra. They build solitary nests of moss and grasses on small, raised hummocks. Their long tails, used in courtship display flights, are shed during the non-breeding season.

← **Many species of terns**, such as these Greater Crested Terns, nest in dense colonies where they can almost touch their neighbors while incubating.

↓ **Black-legged Kittiwakes** nest in large colonies of hundreds or thousands of birds, laying their eggs on narrow cliff ledges. Small ledges make it difficult for aerial predators to land and steal kittiwake eggs or vulnerable chicks.

MATING SYSTEMS

Gulls, terns, skimmers and auks mate for life, although pairs may separate if they are unsuccessful at raising young or cannot coordinate incubation shifts. Male gulls, terns and skimmers feed their females during courtship. Bringing fish back to females presumably leads to larger eggs and higher reproductive success. Smaller species, such as shorebirds and plovers, are usually monogamous, at least for one year. They have a much shorter lifespan than larger members of this order. Phalaropes exhibit sex reversal: females perform courtship, lay their eggs, then leave the male to incubate them while they seek out another male for pairing. Female phalaropes are brightly colored, while males are dull colored and cryptic during incubation. Others species, such as Mountain Plover, can be monogamous, polygamous or polyandrous, depending upon the availability of food.

Pigeons and sandgrouse

Class	Aves
Orders	2
Families	3
Genera	46
Species	327

Pigeons have short bills, necks and legs, and are fairly round and plump. They are strong fliers and their voices are mainly soft, cooing and sometimes repetitive. Pigeons feed on seeds, grain, fruit and invertebrates. Fruit pigeons have jaws with elastic sockets that stretch so wide they can eat fruit larger than their heads. Pigeons are unusual in being able to suck water into their bills. This allows them to drink quickly at desert waterholes. Both males and females produce crop milk, which they feed to their young by sloughing off the lining of their crop. There is a prolonged period of parental care. The familiar pigeon is a descendant of the wild Rock Pigeon, which has declined greatly in its native cliff habitats and is genetically threatened by interbreeding. The terms dove and pigeon are used interchangeably.

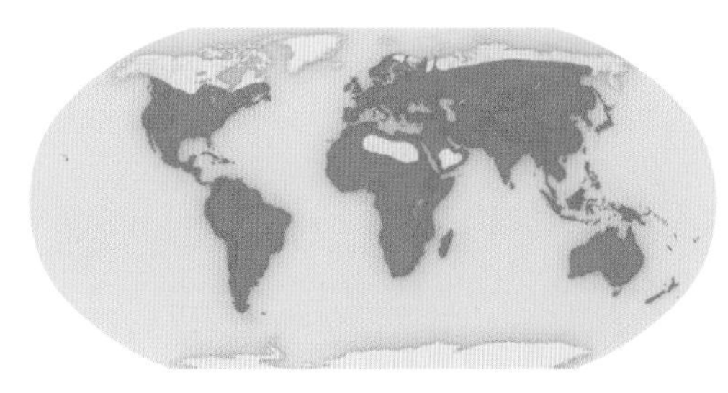

SANDGROUSE

Sandgrouse live in open, treeless, dry habitats in Africa and Eurasia. They survive on seeds in arid areas and fly long distances in large flocks to visit remote waterholes. Although light colored plumage helps them to blend in with the dry sand, they are extensively hunted for their tasty flesh. To cope with the high solar radiation of arid areas, sandgrouse carry water on their belly feathers to wet their eggs. Unlike pigeons, chicks are precocial and leave the nest soon after they hatch. Some species are sedentary, some migratory and some nomadic. When disturbed, they burst upward and fly with quick wing strokes that make a whistling sound.

↑ **Thick-billed Green Pigeons** spend much of their time in trees, where they seem to disappear. Despite blending in, they can easily be heard and have been hunted extensively.

← **The Spinifex Pigeon** is also known as the Plumed Pigeon because of its unusual feathered crest.

↓ **Namaqua Sandgrouse** drink at waterholes in dense flocks at dawn or dusk. Here, in the Kalahari Gemsbok National Park in South Africa, they wet their belly feathers while they drink and carry the water back to moisten their eggs.

Parrots

Class	Aves
Orders	1
Families	1–3
Genera	85
Species	364

Parrots have been loved for centuries and are familiar to people everywhere. Most species of parrots, parakeets, macaws, cockatoos, cockatiels, lories, lorikeets, lovebirds and budgerigars are brightly colored. In size they range from 3.5-inch (9-cm) long pygmy parrots of the Papuan region, to 40-inch (100-cm) long macaws of the Amazon. Lovebirds and Amazons are plump and round, while lories are slender. Some cockatoos have crests that they erect when calling. All parrots have large heads and short necks, with a strongly curved upper mandible that overlaps a shorter mandible. Most are strong fliers, although there are flightless parrots in Australia and New Zealand. Their raucous calls for potential mates or flock mates travel across the forest. Parrots live in a variety of habitats, from deserts to moist tropical forests and alpine grasslands. And they live at all levels in the forest, from the ground to the canopy. Parrots eat mainly fruit, nuts, grains, vegetable matter and some animal matter when they encounter it. Most seed-eating birds disperse seeds, but parrots destroy them during manipulation and metabolize any remains in their digestive tract. They have opposable, or zygodactyl, toes—two face forward and two face backward. This allows them to manipulate objects better than most other birds. Most parrots are left-footed.

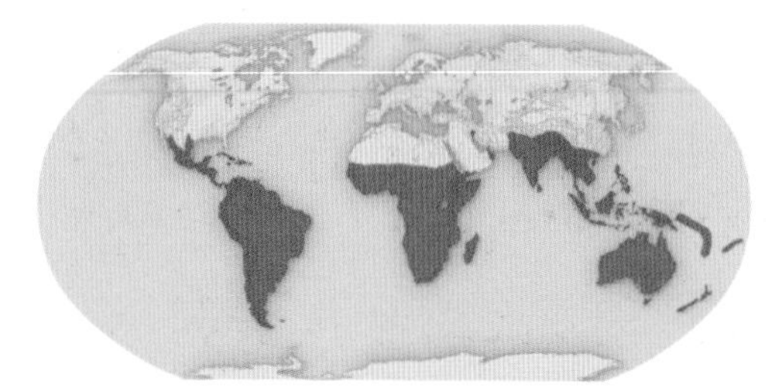

← **White-bellied Parrots** are among many South and Central American parrots that visit clay licks to obtain nutrients from the soil. Where possible, they descend on branches in front of the lick.

↑ **Scarlet Macaws** mate for life, like most parrots, and spend a great deal of time in pair-bond maintenance behaviors, such as mutual preening. Macaws eat large fruits. They require substantial tree cavities for nesting but will adopt artificial boxes placed in the forest canopy. Like many parrots, macaws are victims of an inexhaustible pet trade; however, captive breeding offers hope for survival.

↗ **This Orange-winged Parrot** shows how parrots hold fruit in their feet and peel off the skin. As is typical of parrots, this one is holding the fruit with its left foot. It is a common species that ranges over much of northern South America.

UNUSUAL BEHAVIOR

While most parrots can fly, the Kakapo, also called the Owl Parrot, is a flightless parrot that is mainly active at night. The raptor-like Kea from New Zealand is a large, aggressive parrot that pulls nails out of roofs, rips off windshield wipers and linings of cars, and feeds on dead sheep or other animals it finds. Cockatoos raise their crests of long, pointed feathers as a warning to intruding predators and approaching people, or to attract mates. Pygmy parrots act more like woodpeckers than parrots, creeping along trunks and branches in search of insects. In captivity, as well as in the wild, African Gray Parrots mimic a wide range of bird songs, other animals' calls and any interesting artificial sounds. Although they have not been domesticated, more species have been tamed and kept in captivity than any other group of birds. This is largely because of their bright colors, sociable nature and high levels of intelligence.

Parrots continued

← **Eclectus Parrots from Australasia** are unusual because the males are a bright green with a small patch of red, while the females are a brilliant red with shades of blue. Even their bills are different colors. Parrot fanciers and ornithologists first thought the males and females were two completely different species.

→ **Most parrots nest in holes in trees** for protection from predators. Here, a nearly fledged Blue-and-Yellow Macaw peers out of its nest hole just below its parent.

↓ **Little Corellas** are small, Australian cockatoos. They travel in flocks that can be so dense, the birds look like blossoms on trees. Farmers consider them pests.

↓ **Monk Parakeets** build large communal nests where tens or hundreds of birds nest together, partly to keep warm during the cold winters of southern Argentina. When they nest in palm trees in the Pantanal of Brazil, they usually nest with fewer than ten other pairs. The palm trees are often near houses or barns and provide some protection from other birds, such as hawks, which avoid human habitation.

Cuckoos and turacos

Class	Aves
Orders	2
Families	2
Genera	41
Species	161

Cuckoos, anis, roadrunners, Hoatzins and turacos have a worldwide distribution. Cuckoos and anis live in forests and savannas and eat mainly insects. Roadrunners live in the southwestern deserts of the United States and eat mainly lizards and small snakes. True to their name, they often dash along or across roadways. Some African turacos are known as "go-away-birds" because one species has a call that sounds like "go away." Other turacos give raucous shrieks or grunts. These crested birds eat fruit, seeds and invertebrates, and live in or on the edge of forests. Hoatzins are unique and sometimes placed in a separate order. They live in flocks on lake and river edges in the neotropics; young have claws on their wings to help them clamber through vegetation. Hoatzins ferment leaves in their crop to aid digestion.

BROOD PARASITES

Old World cuckoos lay their eggs in the nests of other species of birds—a practice known as brood parasitism. The female about to lay watches her intended hosts as they build nests, and usually lays her egg in the host nest the same day the female host lays her first egg. The incubation period of the cuckoo is only about 12 to 13 days, much less than that of its hosts. The cuckoo chick hatches first and is usually larger than the host parents' young, giving it a competitive advantage. Sometimes cuckoo chicks lift host eggs or chicks out of the nest and eliminate all competition for food. While some hosts recognize the slightly larger cuckoo egg and push it out of their nest, other species do not.

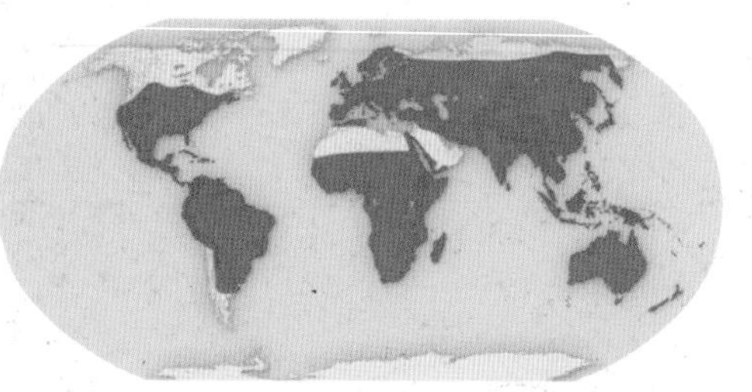

A small Reed Warbler parent is feeding a young, but much larger, begging cuckoo. By this time, none of the warbler's own chicks has survived and the young cuckoo gets the food normally provided for a brood of three or four warbler chicks.

A recently hatched, naked Common Cuckoo chick has the strength to push the eggs of its foster parents, Bull-headed Shrikes, out of their nest.

Turacos, such as this Red-crested Turaco, have brilliant red patches in the wing that contain a unique red pigment (turaxin), which has been used as a dye.

Owls

Class	Aves
Orders	1
Families	2
Genera	29
Species	196

Owls are distinctive and have figured prominently in lore and legend in many cultures. They live in nearly every habitat, including the frozen tundra of the Arctic (Snowy Owls), at the treeline (Hawk Owls), in deserts (Elf Owls), short grasslands (Burrowing Owls), savannas (Ferruginous Pygmy Owls), boreal forests (Boreal Owls), deciduous forests (Great Horned Owls, Screech Owls), and around human habitation (Barn Owls). Not only are owls diverse birds as a group, but individual owl species also vary greatly. Most are nocturnal predators with soft, fluffy plumage for silent flight and excellent hearing to locate prey under the cover of darkness. Their ears are slightly asymmetrical, which allows them to locate the precise direction of sounds and to pinpoint their prey. Most owls primarily feed on small mammals which they capture in their strong talons, then swallow whole. The indigestible parts of their prey, such as bones, skin and hair, are formed into pellets and regurgitated. Ornithologists can study an owl's diet by examining these pellets.

↓ **This Bengal Eagle Owl**, from India, has a typical flat face with forward-facing eyes that are fixed in their sockets. To compensate for fixed eyes, owls are able to rotate their heads in all directions. They turn their heads almost upside down to get another perspective.

Owls continued

NIGHT ADAPTATIONS

Most owls are nocturnal and are adapted to avoid predators during the day, find and capture prey at night, and court and display in the dark. Loud, low-pitched calls, which travel through the forest or countryside, help owls locate mates and defend territories. Their eyes are much larger than they appear, with more rods than cones; rods are sensitive to low light, while cones convey color information. Owls listen for the movements of their prey. They have excellent three-dimensional hearing because their external ear canals are at slightly different levels on their head. Their wing feathers are specialized for silent flight. During the day, owls remain motionless in tree holes, among thick branches or along tree trunks.

← **When they are threatened**, owls, such as this young Great Horned Owl, either fly away or stay to defend themselves. They spread their wings, erect their feathers and hiss loudly. Their large facial disks amplify the appearance of their eyes and, accompanied by outspread wings, are sufficiently threatening to scare away many predators. Female Long Eared Owls defend their nests in this way.

↓ **This sequence is a composite**, high-speed photographic image of a Barn Owl in darkness. It is swooping down to capture prey, guided by sound. Specialized wing slotting that leads to almost silent flight is apparent in the second image. Raising its wings, the owl extends its talons, ready to grasp its target. When the prey is captured, the owl will probably bite it on the neck to finish the job. Owls hold large prey with their feet and tear it apart with their hooked bills before it is eaten.

← **Great Gray Owls**, and many other northern owls, begin incubation late in winter when snow still covers them. Although it looks cold, the temperature of the snow is just at freezing, while the surrounding air may be much colder.

↓ **A close-up of the talons** of a European Eagle Owl shows how strong the owl's toes and claws are for capturing prey. Both the feet and the toes are completely feathered for insulation.

↓↓ **This Long-eared Owl's feather** shows the branched hairs that disrupt turbulent vortices in the airflow over the wing to allow silent flight.

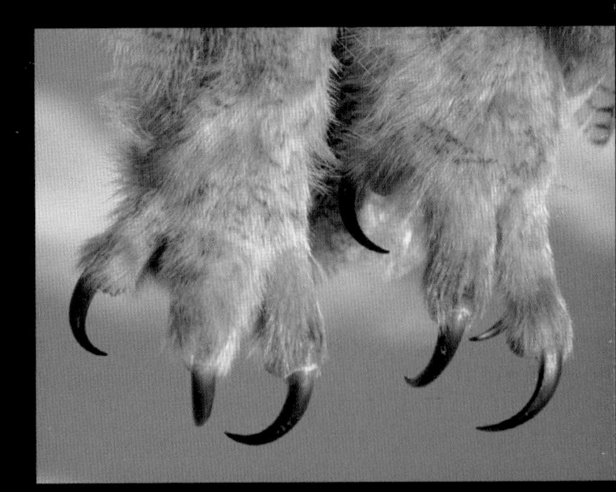

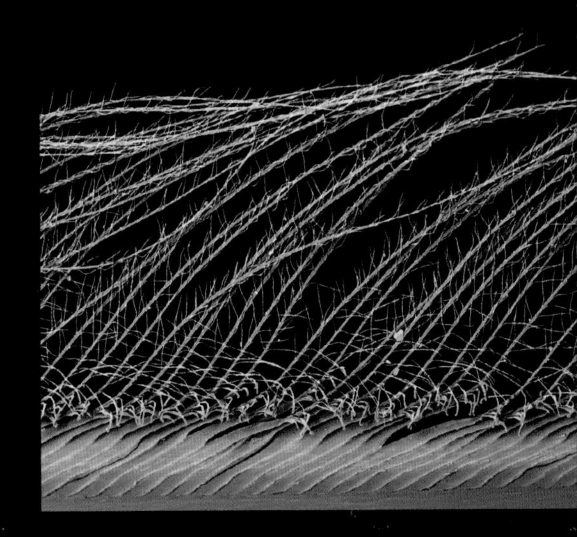

Nightjars and allies

Class	Aves
Orders	1
Families	5
Genera	22
Species	118

As birds of the night, nightjars, poorwills, frogmouths, potoos and Oilbirds are cryptically colored in mottled, barred or streaked browns, grays and blacks. This helps them blend in with tree trunks and the ground while they sleep during the day. Many nightjars nest on the ground. Like owls, they have soft plumage for quiet flight and excellent vision and hearing. They sally from perches or feed in flight, opening their wide mouths to scoop in flying insects. Larger species sometimes catch small birds. Nightjars are named for their loud, persistent and monotonous calls that "jar" the night. Their short legs and small feet are almost useless for walking; nightjars shuffle along the ground. Poorwills are unusual in the group in their ability to go into torpor. When air temperatures reach 66° F (19°C), they hide in rock crevices and reduce their own body temperature to 64°F (18°C).

CURIOUS RELATIONS

Frogmouths, potoos and oilbirds all behave differently from nightjars. Instead of catching insects on the wing, frogmouths perch on a stump, peer at the ground, then swoop down on invertebrates and mice. In contrast to most nightjars, they nest on tree branches and camouflage their nests with leaves, moss and lichens. Potoos hunt more like flycatchers; they sit on a perch and sally forth to catch flying insects. Oilbirds are nocturnal. They nest in colonies in caves and feed on fruits; nests are made from their own droppings. Fledglings can weigh twice as much as their parents.

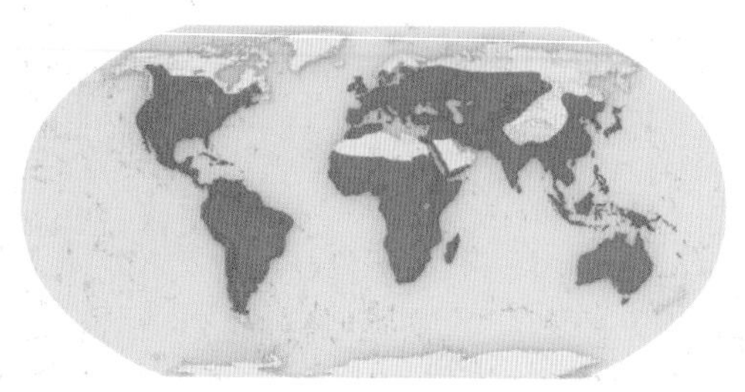

↑ **Frogmouths, such as the Tawny Frogmouth,** sometimes roost on tree limbs where they look like a dead snag. Would-be predators mistake them for an extension of the tree.

← **Oilbirds fly at night** using echolocation to navigate in their caves. Here, an Oilbird broods its gray chicks at a nest site in Trinidad.

Hummingbirds and swifts

Class	Aves
Orders	1
Families	3
Genera	124
Species	429

Hummingbirds and swifts are generally small, with short, weak legs and small feet. Swifts feed only on flying insects, while hummingbirds eat nectar, insects and spiders. Swifts are fast and elegant fliers. They soar over marshes and grasslands, and some even mate in flight. Several species can fly at speeds exceeding 100 miles per hour (160 km/h). Crested swifts are a unique group. They have crests and metallic colors, and occasionally perch in trees. The saliva that some swiftlets use to glue their nests together provides the basic stock for birds' nest soup, a popular dish in Asia.

Hummingbirds frequently perch on flower stalks or small branches. They use their long, thin bills to suck up nectar by capillary action. Hummingbirds are important pollinators for many plants. Their most striking features are their diminutive size; bright, iridescent colors; and rapidly whirling wings that allow them to move in all directions or hover at flowers.

↗ **Strong, agile fliers**, Great Dusky Swifts fly near the crashing water at Iguazu Falls in Brazil. They nest behind the falls, where they are safe from predators.

→→ **Seychelles Cave Swiftlets** nest in colonies on walls. They use saliva to glue their nests to the wall.

→ **Young Common Swifts** can cling to any vertical surface, and hang by their feet on the sides of walls.

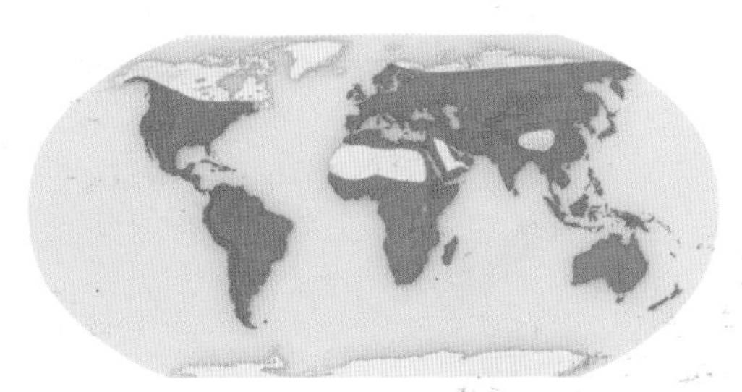

Hummingbirds and swifts continued

SMALL BUT SPECTACULAR

Although small in size, hummingbirds are among the most spectacularly colored birds. Their dazzling plumage has a brilliance and iridescence unsurpassed in other birds, and in the nineteenth century their feathers were used in adornments such as brooches and pins. The feather's iridescence depends on the angle at which sunlight strikes it, often changing from black to a brilliant red, purple or green. For their size, hummingbirds have a high metabolic rate, which provides them with high energy levels for hovering. A concentration of muscle mitochondria allows rapid energy output, and even when they are perched, their metabolic rate is higher than that of other birds. When night temperatures drop, hummingbirds go into torpor—by lowering their body temperatures they are able to conserve energy. They aggressively defend their territories and flower patches.

→ **The Scintillant Hummingbird**, shown here feeding on and pollinating an epiphytic heath in the cloud forest of Costa Rica, has a straight bill for small flowers.

↓ **The arrangement of the feathers** and wing bones in hummingbirds is designed to transfer enormous power to the flight feathers. Unlike other birds, the shoulder bones are attached to the shoulder girdle so that the wing can swivel. This allows them to hover for brief periods and fly in almost any direction, forward or backward, up or down. The tremendous wing muscles average between 25 and 30 percent of a hummingbird's body weight.

HUMMINGBIRD WING STRUCTURE

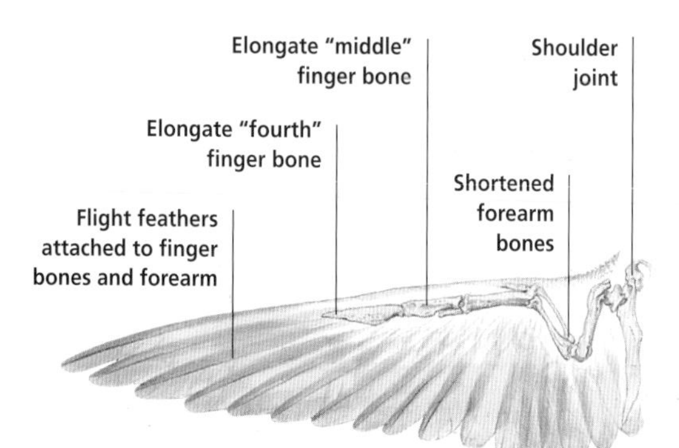

← **In the early morning** Sparkling Violet-ear Hummingbirds dip into flower leaves and bracts to sip water or dew, as well as to obtain nectar. The size and shape of a hummingbird's bill are specialized to its nectar flowers; bills are short or long, straight or curved, depending on the shape of the flowers.

← **An Anna's Hummingbird** sits on the edge of its nest, made of moss, leaves, spiderwebs and other bits of vegetation, feeding its young. In most species, only the female incubates the eggs, which are placed in a small cup nest. Chicks beg when they feel the breeze from the female's approaching wings and are fed insects as well as nectar.

↞ **Slow motion photography** of a Sparkling Violet-ear Hummingbird in flight shows its ability to swivel its wings at different angles.

Mousebirds

Class	Aves
Orders	1
Families	1
Genera	2
Species	6

Mousebirds, also known as colies, live in the forests and savannas of Africa. They are named mousebirds because they scurry through vegetation like mice. They eat fruit, plants and some insects, and travel in chattering flocks of 20 or more birds, moving from bush to bush. Farmers and gardeners dislike them—mousebirds destroy berries, strip off leaves, eat flower buds and peck holes in fruit. They also prey on the nestlings of other birds and so are often mobbed by noisy flocks. Both sexes tend their two to four eggs in a tree nest of grasses, leaves, bark and roots. The naked chicks are fed partly digested food by regurgitation. They leave the nest within a few days of hatching to creep about the branches.

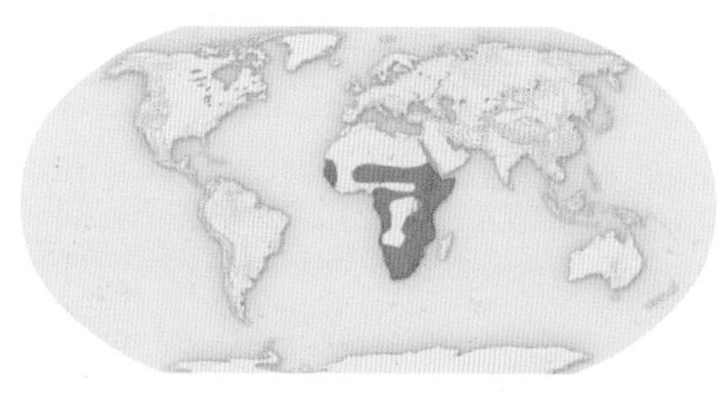

A Brown Mousebird sits in the fork of a branch in Kenya, with its flock mates not far away. Like other mousebirds, it is somber colored but has bright red feet. The toes of a mousebird can splay in all directions.

Trogons

Class	Aves
Orders	1
Families	1
Genera	6
Species	39

Trogons are brightly colored, sedate and solitary inhabitants of tropical forests. They have two toes that point forward and two pointing backward, a trait shared only with parrots, woodpeckers, cuckoos and toucans. Most trogons have long, square tails and green plumage on their backs. Many have brilliant red or yellow underparts. The unique red pigment eventually breaks down in study specimens and the feathers appear pink. Trogons eat fruit, large insects, snails, lizards and frogs. They nest in holes in trees and stumps, which they often excavate themselves; many other species nest in the holes trogons make. The most ornate trogon is the Resplendent Quetzal. Trogons are non-migratory.

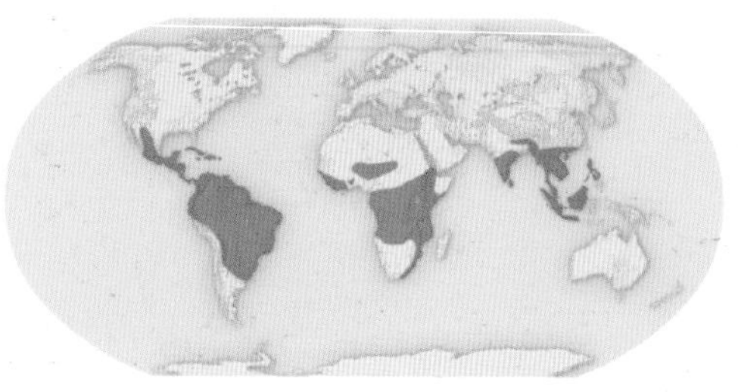

Blue-crowned Trogons, like other trogons, often perch on cross branches to give their short, whistled song. They are difficult to locate in the dense undergrowth of forests because they can sit for hours without moving.

Kingfishers and allies

Class	Aves
Orders	1
Families	11
Genera	51
Species	209

Most kingfishers and their allies are brightly patterned or dramatically colored and have small feet with three fused, forward toes. They include birds such as kingfishers, rollers, motmots, bee-eaters, hornbills and hoopoes. Birds in this order vary in size, from 3.5-inch (9-cm) long West Indian todies, to 5-foot (1.5-m) long hornbills. Some species, such as bee-eaters, are highly colonial while others, such as kingfishers and hornbills, nest solitarily. All are carnivorous, devouring fish, reptiles, amphibians and small mammals, and a few species eat fruit and berries as well. Kingfishers have large, robust bills that are useful for catching fish, while rollers and motmots have slightly hooked bills for holding and crushing prey or grasping fruit. Although bee-eaters feed on many kinds of insects, over half of their diet is made up of bees. They snatch bees and other stinging insects in midair, then return to a perch, wipe off the sting and swallow them. Only kingfishers occur worldwide, nesting everywhere except in polar regions. The graceful bee-eaters range throughout most of temperate and tropical Europe, Africa and Asia; the temperate species are migratory.

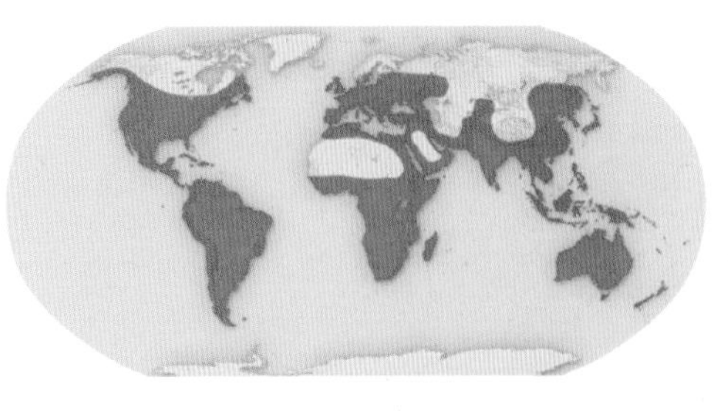

A Common Kingfisher with a fish sits quietly on a dead tree stump. It will toss its head vigorously to position the fish so it can be swallowed headfirst. Some kingfishers are excellent fishers, while others forage in trees, shrubs, or on the ground.

Kingfishers and allies continued

PROTECTIVE NESTS

Nearly all kingfishers and their relatives have bold color patterns, many with patches of red, black or white. Such coloration, normally used to attract mates or other members of the species, can make birds vulnerable to predators. In response to this, most kingfishers and allies nest in holes in banks or in dead trees, where it is difficult for predators to access eggs, chicks and incubating adults. Some species, such as bee-eaters, nest in dense colonies where members warn each other of approaching predators and actively mob any that come too close. Some hornbills carry nest protection to an extreme: the female is sealed in a tree-cavity nest during incubation. Both male and female work to plaster up the tree hole with mud, leaving only a small slit for the female to stick out her bill so the male can feed her.

↑ **Male Eurasian Bee-eaters** court females by presenting them with small fish or insects. Males provision females with additional food to cement the pair bond and ensure high egg quality.

↑ **Like all rollers**, Lilac-breasted Rollers of eastern and southern Africa are noted for their "rolling" aerial acrobatics. They perch conspicuously on branches waiting for an unsuspecting insect, then sally forth to seize it. Rollers are stout birds with thickened bills and patches of iridescent blue in their plumage. They are particularly vocal at the onset of breeding.

↑ **Kookaburras** are Australian forest kingfishers known for their laughing cries, which they often give in the early morning and again at dusk. They feed on snakes and lizards, as well as the young of other birds.

↑ **Ground Hornbills** are large, black birds that spend their time ambling along the ground in search of insects, lizards and frogs. When threatened, they run quickly or take flight suddenly, flashing conspicuous white wing patches.

Woodpeckers and allies

Class	Aves
Orders	1
Families	5
Genera	68
Species	398

Most woodpeckers, jacamars, puffbirds, honeyguides, toucans and aracaris nest in tree or mound cavities and lay white eggs. They have zygodactyl feet, with two toes pointing forward and two back, which allows them to grip vertical tree trunks. Woodpeckers are small to medium-sized, mostly arboreal birds with stout, chisel-like bills for probing in wood. Stiff tail feathers, or rectrices, allow them to use their tails to brace themselves while clinging upright on tree trunks. Among woodpeckers, several species of flickers are mainly terrestrial, nesting in holes in the ground and feeding on ants. Many other species of birds rely on the abandoned nest cavities of woodpeckers for their own nests. Several woodpecker species have adapted to urban and suburban life, often drumming on telephone relay boxes, roofs and other metallic objects to attract females and ward off competitors. Barbets have either a stout bill for eating fruit or a slender bill for catching insects. They are small, plump birds of lush forests, and are usually brightly colored with gaudy red, yellow and green plumage. Jacamars are small to medium-sized birds with long, slender bills for catching aerial insects, including butterflies. They are graceful birds of the forest and are closely related to the puffbirds. Most have iridescent green back plumage. Toucans are intermediate to large fruit-eating birds with enormous, brightly colored bills. The drab honeyguides are the most unusual in the group because, like cuckoos, they are brood parasites.

BRIGHT PLUMAGE CAN BLEND

Many species of woodpeckers, toucans, aracaris and jacamars are brightly colored. The dramatic black and white pattern of woodpeckers blends in with their tree trunk habitat, particularly when they are still. When jacamars remain motionless, their brilliant and often iridescent plumage mixes in with forest foliage, mimicking flowers and fruit. Toucans seem particularly obvious because of their large and colorful bills, but they, too, blend in when they sit still. Surprisingly, brightly colored birds can be difficult to see because patches of color break up the lines of the bird, so its outline is difficult to recognize. Instead, the eye notices a splash of color, which could be a flower or fruit, rather than the whole shape of a bird.

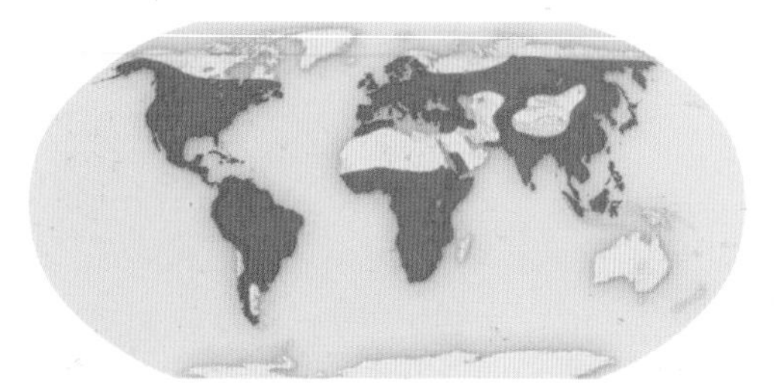

← **This Great Spotted Woodpecker** is storing an acorn in a tree. Woodpeckers use their strong bills to probe for insects, pierce nuts and seeds, and dig nest cavities.

↑ **Greater Honeyguides**, like all honeyguides, are dull colored birds that eat insects and wax from beehives. In Africa, they are sometimes called "indicator birds" because native Africans follow them to hives.

↗ **A Plate-billed Mountain Toucan**, perched on a tree in the Andes Mountains of Ecuador, holds fruit in its large bill. Although a toucan's bill can be a third or more of its entire body length, the bill is surprisingly light. Its interior is made up of a network of bony fibers and is not solid.

→ **This Red-breasted Sapsucker** peers out of a nesting cavity in California, holding wood shavings it is removing. Sapsuckers drill holes in wood to suck out the sap, leaving a ring of holes around trees where they have been foraging.

Passerines

Class	Aves
Orders	1
Families	96
Genera	1,218
Species	5,754

Passerines are perching birds. They include about 60 percent of the world's birds and are by far the largest bird order. They have proven to be remarkably adaptable and are the dominant land birds on all continents except Antarctica. Passerines have exploited more niches in more habitats than any other order of birds. They range in size from the 4–6-inch (10–15-cm) kinglets, warblers and sparrows, to 26-inch (54-cm) long ravens. While the number of wing and tail feathers varies in non-passerines, all passerines have 12 feathers in their tail and 9 or 10 primary wing feathers. They feed on fruit, seeds, nuts, other plant material and insects; some species specialize in diet while others are generalists and eat a variety of foods. Most, however, feed insects to their young because insects are more easily digestible than fruit and provide more protein. Although there is some variation in mating systems among passerines, most are monogamous and both parents raise the young. Passerine chicks are altricial, born naked, and require parental care and protection. As an order, they are most noted for their calls and songs, which can be incredibly varied and beautiful. They sing to communicate and to mark their territory. Poets have been inspired by the songs of species such as the Skylark, Hermit Thrush and Pied Butcherbird.

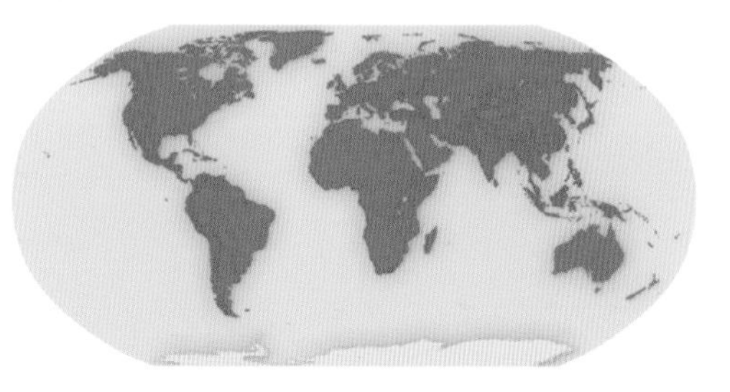

→ **The song of the Skylark**, one of the most melodious in the world, was celebrated in Shelley's poem, "Ode to a Skylark." Skylarks rely on their song, rather than colorful plumage, to attract mates.

↘ **Western Meadowlarks** perch on grass seed heads and cattails to give their territorial display calls.

↙ **The Hermit Thrush**, perching here on a small stump, has a long, liquid warble that drifts through the forest in the dawn chorus. It is almost flutelike.

PASSERINE TOES

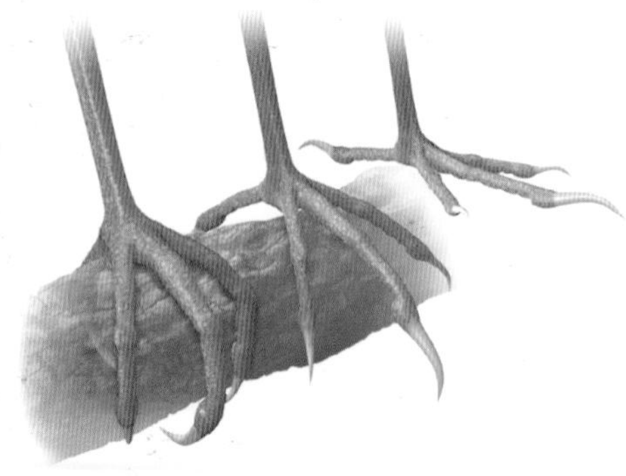

PERCHING BIRDS

The feet of passerines are well developed for perching on limbs or twigs. Three of their toes point forward and the other enlarged toe, the hallux, with a long claw for grasping, faces backward. The toes all join the leg at the same level, which gives maximum flexibility in perching and grasping small objects. The muscles and tendons in the leg are arranged so that if the bird starts to fall backward, the muscles automatically tighten the grip and lock the bird in place. Even at rest, their claws grasp the perch. The simplified muscles are, however, at the expense of more delicate toe movement. Passerines cannot manipulate objects as well as birds in some other orders.

Passerines radiation

PASSERINE ORIGINS

Fossil evidence and molecular techniques have helped scientists understand patterns of passerine evolution and radiation. In the past, ornithologists believed that passerines evolved in Eurasia 30 to 20 million years ago, then spread to the rest of the world. New fossil evidence from Australia and New Zealand indicates that passerine radiation is much older, and occurred in the Southern Hemisphere, on Gondwana, some 85 to 80 million years ago. Two major radiations occurred. Oscines and suboscines in Gondwana produced lines that moved first into the Southern Hemisphere continents, then northward. Once the oscines reached Eurasia, there was a second burst of radiation with species moving into North America from England, Greenland and northern China.

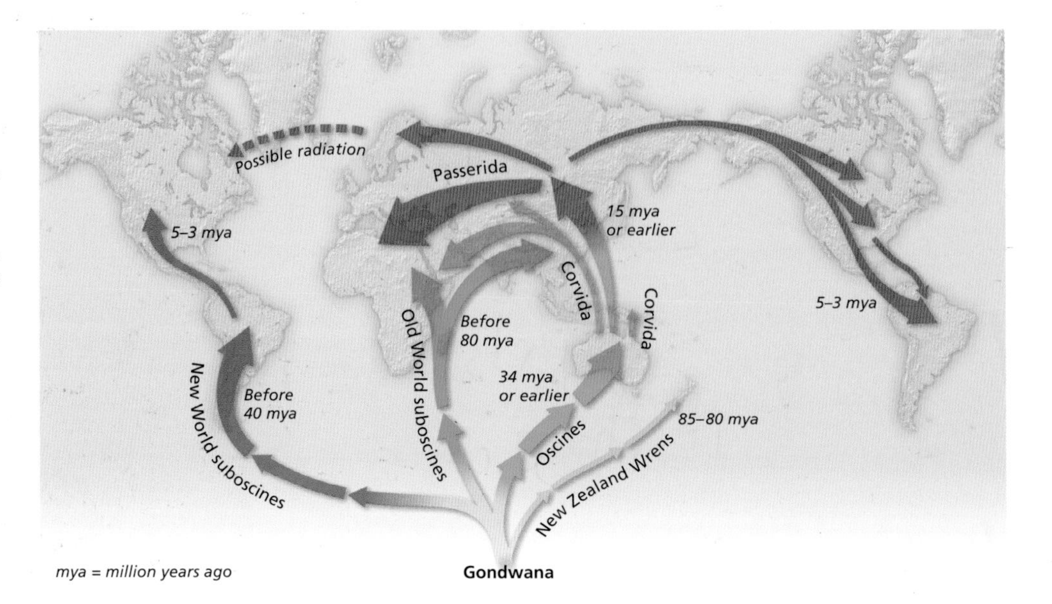

↓ **Vermillion Flycatchers** came from the passerine radiation that moved into the New World through South America. This is a male; females are pale gray.

The Rifleman is one of two small surviving New Zealand wrens. These birds have their own family that radiated into New Zealand some 85 million years ago. The Rifleman is fairly common. It has a short tail and feeds on insects in mature rain forests.

Flame Robins live in open, grassy country. They belong to a group that radiated into Australia at least 35 million years ago. They feed on grubs and worms on the ground.

Banded Pittas live in Southeast Asia and Indonesia. They came from a radiation that occurred more than 80 million years ago. Pittas live on the forest floor where their dark coloration and belly bands help camouflage them. Their long legs allow them to hop rapidly through leaf litter in search of insects.

Indigo Buntings, American birds that appear blue all over, represent a recent radiation of buntings into North America—only 5 million years ago. Buntings forage in trees, in shrubs and on the ground, eating insects, seeds and fruits.

Passerines diets

FORAGING AND DIET

Perching birds use a wide range of foraging methods and have varied diets, including nectar, fruits, seeds, acorns, other nuts, other vegetation, insects, small vertebrates' eggs and carrion. Their perching ability allows them to exploit a diversity of feeding substrates because they can perch on small stems and branches, and walk or hop along the ground. Passerines hover to search for insects on tree bark (tits and chickadees); probe in bark (treecreepers, nuthatches and tits); search small branches (warblers); and sally forth from branches to catch insects (flycatchers). They ride on other animals to pluck insects off their backs (oxpeckers); creep through the grass to garner insects and seeds (grass wrens and sparrows); pull worms and grubs from the ground (robins and thrushes); and snatch insects scared up by swarming ants (antwrens and antpittas). Others sip nectar (sunbirds, some chats and honeyeaters); catch insects on the wing (swallows); pick seeds off delicate grass stems (finches); and scurry along the ground to take small animals (lyrebirds). And these are only some foraging methods. Dippers are the only passerines that are aquatic. They use their wings to swim underwater in search of insects.

→ **Malachite Sunbirds** sip nectar from a range of flowers, such as South African protea and rat's tails. They forage alone and occasionally eat insects.

↓ **Eurasian Nutcrackers** live in coniferous forests in the mountains of central Europe. They specialize on hazel nuts. In the non-breeding season they travel in small flocks and sometimes invade eastern and western Europe when the local nut crop fails.

← **Sunbirds are nectar-feeding birds** that are the Old World equivalent of hummingbirds. Some are brightly colored, others drab. Here, one perches on the closed blossom of a banana plant in Borneo.

↓ **Carrion Crows are omnivores** and eat nearly everything. They frequently eat the eggs of other birds, often working together to draw the incubating parent away from the nest. Crows and jays are the most intelligent passerines.

↓↓ **The House Wren**, shown here with a small cricket, mainly eats small insects, such as grasshoppers and caterpillars. It occurs in a wide variety of habitats, from towns to forest edges, at all elevations, from Canada to Tierra del Fuego.

Passerines classification

SUBOSCINES AND OSCINES

The classification of birds reflects radiation and evolutionary relationships among different groups. Passerines are the most problematic of all the bird orders to classify because of their rapid diversification into more than 5000 species, and the potential for convergence—where birds evolved similar structures or behaviors to solve the same problem in different places in the world. Modern molecular techniques, which often involve examination of DNA, can elucidate relationships among closely related birds. But there is still no agreement on how to arrange the passerines. Not all passerines are songbirds. There are more than 80 families within the order, which is divided into two suborders: the suboscines and the oscines. Ornithologists make the distinction between songbird passerines and non-songbird passerines based on their syrinx—the unique structure at the base of the windpipe that produces sound. The suboscines have a relatively simple syrinx, while the oscines have a complex syrinx and are the "songbirds." Suboscines, such as ovenbirds, antbirds, woodcreepers and tapaculos, make simpler vocalizations than oscines. These calls are not technically recognized as "songs."

← **Winter Wrens** are oscines. Wrens have been divided into more subspecies than any other passerine and have adapted to a range of habitats, including gardens in northern Europe, boreal forests of Canada and island coasts in the Bering Sea.

→ **Swallows are oscines** that have taken to the air, catching insects on the wing. Barn Swallows have adapted to urban environments and now nest in old buildings, on window ledges, under house eaves and bridges, as well as in caves.

→ **Barred Antshrikes are suboscines** that live in the neotropics. They use their sharply hooked bills to tear apart insects. Males, like this one, have striking black and white patterned plumage to attract females and repel other males. Females have brown plumage that provides good camouflage for foraging in the forest.

← **Titmice, such as this Blue Titmouse**, are oscines. They belong to a group of more than 60 species of titmice and chickadees in the Northern Hemisphere. Titmice are incredibly adaptable: for example, in England they are known to remove the foil caps of milk bottles left on porches.

Passerines suboscines

↖ **Brown-billed Scythebills** are members of the woodcreeper family that ranges from Mexico to Argentina. They nest in tree cavities and use their long bills to search bark crannies, probe moss and ferns, and pull grubs and insects from burrows. Their mottled brown plumage is good camouflage.

← **Rufous Horneros** are one of the most common birds of the Argentine and Brazilian pampas. They build round mud nests on trees, posts and abandoned houses. The nest opening curves upward and back to the nest cup, making it difficult for predators to reach the eggs and chicks inside.

↞ **Broadbills, such as this Green Broadbill**, are brightly colored and live in the wet jungles and cloud forests of the Old World. Green Broadbills feed on fruit and berries, although other broadbills eat mainly insects, lizards and a few frogs.

SUBOSCINES DIVERSIFY

The suboscines, in the suborder Tyranni, include birds that diverged into Southeast Asia (broadbills and pittas), the New World (flycatchers, becards, ovenbirds, woodcreepers, antbirds and manakins), and New Zealand (New Zealand Wrens). Their highest diversity is in the New World tropics, especially in rain forests; they make up nearly a third of the birds in the neotropics. Tyrant flycatchers, with more than 400 species, are the largest suboscine family. They have evolved to fit a number of different niches in rain forests, savannas and treeless paramo. Tyrant flycatchers forage in the canopy, middlestory, understory, on the ground and at the forest edge. Woodcreepers have also diversified, although within the much narrower range of foraging on trees. They resemble the Brown Creepers of North America and the tree creepers of Eurasia in color patterns, bill shapes and behavior, which reflects convergence. The different bills of woodcreeper species mirror subtle differences in foraging methods and the location of prey. Antbirds and antpittas live in the forest understory. Some follow army ants and others forage on their own; many antpittas are somber colored. They whistle or hoot loudly to keep in contact with each other as they skulk through thick, impenetrable vegetation. New Zealand Wrens are tiny, insect-eating birds. They are weak fliers.

↑ **A pair of Striped Manakins** sits quietly on a branch in Brazil—the male has the red head. In most manakins, males are brightly colored. They engage in elaborate courtship dances and make sharp, cracking noises that sound like sticks snapping. Some display on their solitary territory, while others have leks where several males display simultaneously.

Passerines oscines

TRUE SONGBIRDS

Oscines that belong to the suborder Passeri are considered true songbirds. They account for roughly 40 percent of the world's birds and can be divided into two main groups: the corvida and the passerida. The corvida, or crow relatives, include lyrebirds, bowerbirds, fairywrens, logrunners, shrikes, vireos, crows, mapgies, jays, nutcrackers, cuckoo-shrikes, drongos and Old World orioles. The passerida make up the rest of the songbirds. Songbirds can usually be identified by their songs, although some birds are skillful at mimicking the songs of other species. Oscines use their songs to establish territories, attract mates and repel competitors for their mates or their territory. Mates, as well as neighboring birds, are able to recognize each other's calls. Members of the corvida are shown on this page, as well as dippers, which are related to the thrushes. Crows and their relatives are among some of the smartest birds. They are often social and some are predators on the eggs and young of other birds. Many engage in elaborate displays with their tail feathers (birds-of-paradise) or by building bowers (bowerbirds).

→ **This male Superb Lyrebird** is displaying in a eucalypt forest in Australia. Its tail is flung over its head and its lacy feathers are spread between two outer tail feathers, which are almost 2 feet (60 cm) long. Elaborate displays such as this are usually accompanied by a loud, musical medley of mimicry and song.

↓ **The five species of dippers**, such as this Eurasian Dipper, are the only aquatic passerines. They spend most of their time hopping from rock to rock amid rushing water, searching for insects and occasionally fish. Dippers have short wings for swimming and long, sturdy feet to grip the stream bed.

↓ **The Raggiana Bird-of-paradise** is common in New Guinea, even in towns. It is depicted on stamps, posters and travel brochures. But Raggiana males, like many other bird-of-paradise species, are rarely seen in the wild due to excessive hunting for their feathers. These birds give noisy displays, flaunting filamentous, scarlet plumes.

Passerines oscines

→ **Arrow-marked Babblers**, like most babblers, are highly social and travel in large family groups. They often have a sentinel who remains in the tops of trees to warn family members of approaching predators.

↓ **Flycatchers, such as the Pied Flycatcher**, primarily catch insects. They sit and wait for an insect to fly by, then quickly sally forth to catch it. They return immediately to a perch and consume their prey. Flycatchers make delicate cup nests from shredded leaves, moss and lichens. They place their nests on tree branches and both parents incubate and care for the young.

OLD WORLD INSECT-EATERS

The Old World insect-eaters include swallows, chickadees, titmice, nuthatches, creepers, bulbuls, kinglets and Old World Warblers. Titmice and chickadees are birds of small woodlots, forests and gardens in the Northern Hemisphere and Africa. They often feed at bird feeders during winter and cache seeds in tree crevices, regularly stealing cached seeds from each other. While titmice and chickadees normally move up trees to forage, nuthatches move down trees, headfirst. Most titmice, chickadees and nuthatches nest in tree holes and often use the abandoned nests of woodpeckers. Bulbuls live in Southeast Asia, Africa and Madagascar. Babblers are the forest-dwellers; swallows are the aerial members of this group.

← **Penduline Tits** are small birds of brush and thickets along streams. Their pendulous nest has a funnel-like entrance that makes it difficult for predators to invade.

↓ **Bulbuls feed on fruit**, although they also eat insects and other small animals. Most are loud and gregarious and some have adapted to villages and small towns.

Passerines oscines

PASSERIDA

This group includes weavers, honeycreepers, larks, wagtails, flowerpeckers, sunbirds, tanagers, sparrows, buntings, New World blackbirds, orioles and finches. This is a large and diverse group, with a range of foraging methods and foods. Blackbirds, orioles and some tanagers feed on insects; weavers, finches and sparrows eat seeds; flowerpeckers and sunbirds sip nectar; tanagers and some honeycreepers specialize on fruit and berries. Most often, however, these oscines eat a combination of foods. Blackbirds and buntings live and forage in grasslands; sunbirds and weavers inhabit open grasslands; while tanagers are open or dense forest-dwellers. Many others move between habitats. The Old World seed-eaters include goldfinches, waxbills and weaverfinches, as well as African weavers. Most weavers are gregarious and travel in flocks of hundreds or thousands, nesting in large colonies. Buffalo Weaverbirds build communal nests that can grow so heavy that the branches collapse around them and send nests to the ground. Queleas are nomadic, waiting for the rains before they nest. Weavers are noisy; they chatter and chirp almost continuously. Goldfinches and their relatives, including redpolls, crossbills and bullfinches, all have undulating flight displays with characteristic songs.

← **Painted Buntings** are one of the most colorful and exotic-looking species of North America. They haunt brushy places and hide their nests in low, dense foliage.

→ **In spring migration, Magnolia Warblers** flit from tree to tree searching branches and blossoms for insects. They nest in coniferous forests of North America, where they glean insects from twigs and needles.

← **Yellow-billed Cardinals** are common and conspicuous songbirds. They occupy roadside and streamside brushy vegetation on the Brazilian Pantanal. These birds are often trapped by villagers as pets or sold for the commercial pet trade.

Distribution and habitat

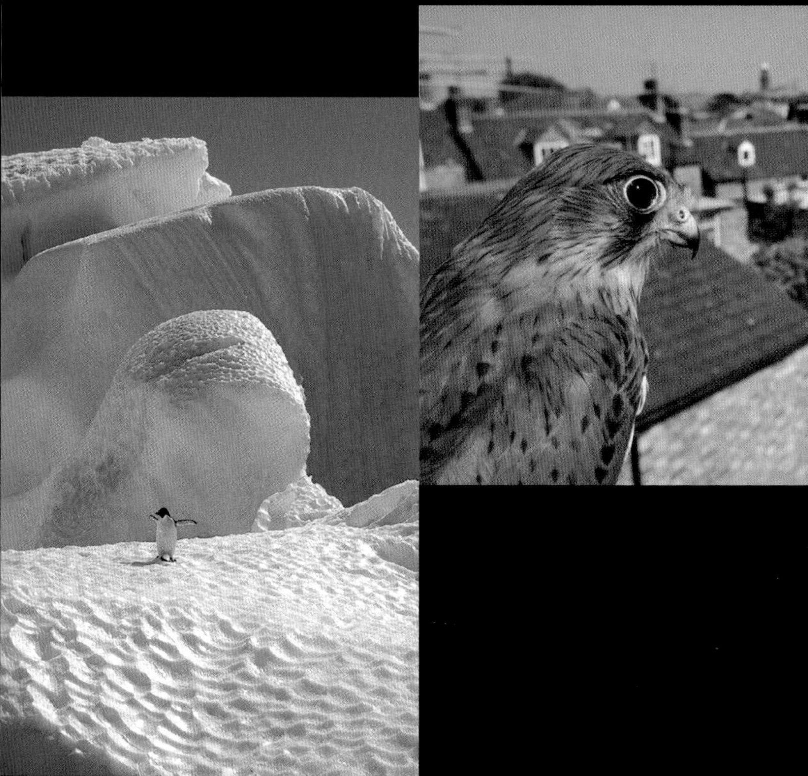

Distribution and habitat

Birds have adapted to habitats as varied as the Arctic tundra, Sahara Desert, Amazon rain forest and the open ocean. While some birds exist within only one habitat, others exploit or regularly move between more than one. Many migratory birds occupy different habitats in breeding and non-breeding seasons.

World zones

The world can be divided into six general regions: the Palearctic, Nearctic, Neotropical, Afrotropical, Oriental and Australasian. Many bird families occur in three or more of these regions. The largest zone, the Palearctic, includes the bulk of Eurasia and northern Africa. Most of its species are migratory. Prominent among its groups are warblers, finches, owls, flycatchers, larks and wagtails. The Nearctic region, which stretches from North America to Mexican rain forests, also has a large proportion of migratory birds. The Nearctic and Palearctic share many bird families, such as grouse, cranes, larks, pipits and finches. The Neotropical region, which includes South America, the West Indies and Central America north to the forests of Mexico, is home to the highest number of bird species. The Afrotropical region encompasses Madagascar, southern Arabia and Africa south of the Sahara. It has high species diversity and shares a number of families, including bustards, broadbills, bulbuls and sunbirds, with the Oriental region. The Oriental region takes in all of Asia south and east of the Himalayas, southern China, Indonesia and the Philippines. Pheasants are particularly well represented in the Oriental region. The Australasian region comprises mainly the Moluccas, New Guinea, Australia, New Zealand and some small mid-Pacific islands. It contains a relatively small number, but a rich diversity, of species.

ENDEMIC SPECIES

Some species of birds are restricted in their range to very small geographical areas, such as isolated islands or remote mountain ranges, or can survive only in a specific and particularly limited kind of habitat. These are said to be "endemic" species. Because of their relative isolation, or the often precarious nature of their habitat, endemic species can be vulnerable to extinction. Some parts of the world have a disproportionately high number of endemic birds, and many species and families of birds occur only within one of the world's six zones. Accentors, for example, are endemic to the Palearctic region; Emus occur only in Australasia. The island of Madagascar has no fewer than five endemic bird families. In the Neotropic region, about one-third of all species are endemic.

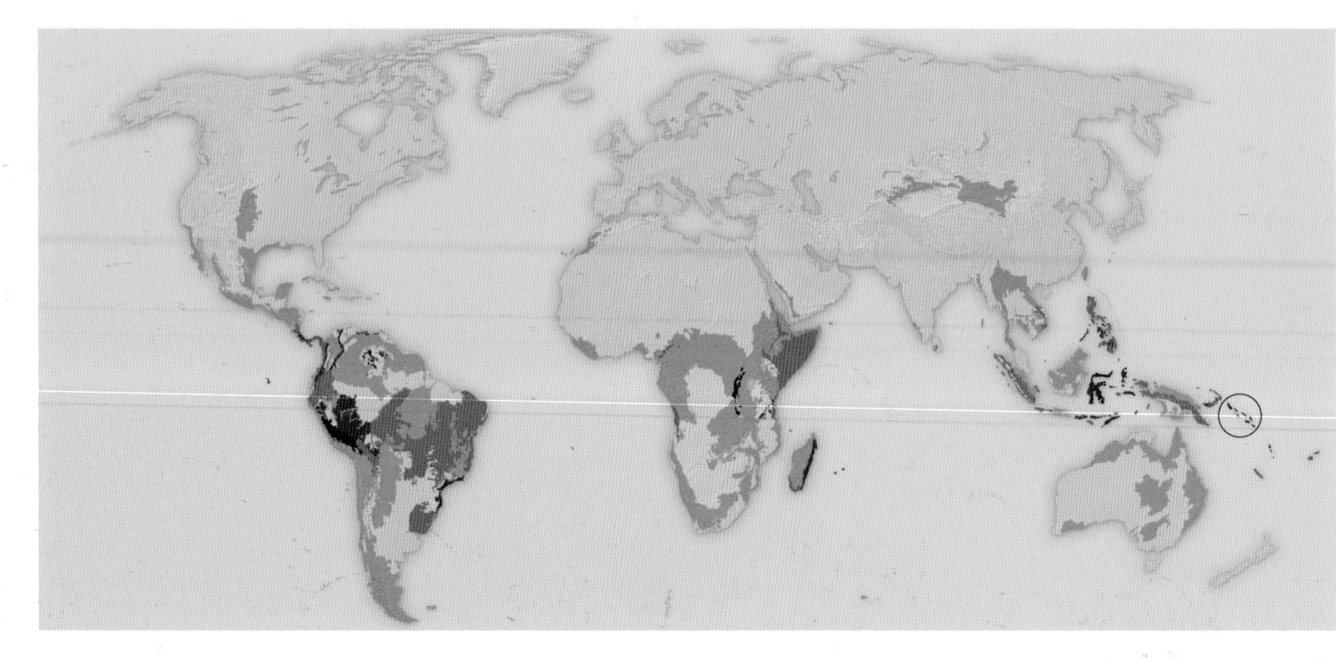

The number of endemic bird species per area is based on information from the World Wildlife Fund. Areas with high levels of endemism warrant critical conservation attention.

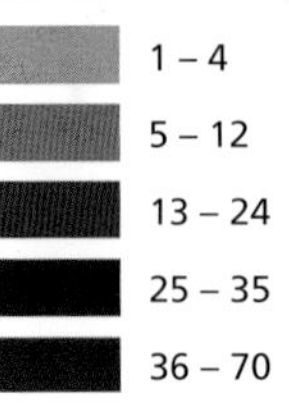

↖ **The Andes are home** to numerous endemic species, including endangered birds such as the Black-breasted Puffleg Hummingbird.

↑ **The Atlas Mountains**, in Morocco and Algeria, are host to a number of isolated, endemic species and subspecies.

← **The King Bird-of-paradise**, the smallest of the birds-of paradise, is endemic to New Guinea, although it is fairly widespread there.

→ **Satin Bowerbirds** are endemic to Australia. Males build bowers to attract females.

Birds with wide distribution

Although, at least theoretically, birds can fly anywhere, they cannot live in inhospitable environments or in places that do not provide their food and habitat needs. The range of any particular bird can be defined as the total area in which it can live and survive throughout the seasons of the year. Most birds are limited to one or two of the world's six major zones, or even to specific locations or habitats within a particular zone. Few birds have worldwide distributions. The key to a wide distribution is flexibility. The range of any particular species depends on its habitat, nesting and feeding requirements, as well as the bird's capacity to coexist with other animals, including humans. Particularly diverse environments provide opportunities for a range of bird species with quite specific and differing needs. A hawk or Osprey that feeds on a wide variety of fish can forage over all the world's oceans, as long as it can find suitable nesting sites and materials. In some cases, a nesting site may be as simple as a narrow pole or tree that protects the bird from mammalian predators.

↗ **Sanderlings** are widely distributed. They breed only in northern temperate and Arctic regions, but fly to nearly every coast when they migrate.

→ **Barn Owls** are among the few bird species with a worldwide distribution. They are flexible in their foraging behavior and habitat selection.

Barn Owl distribution

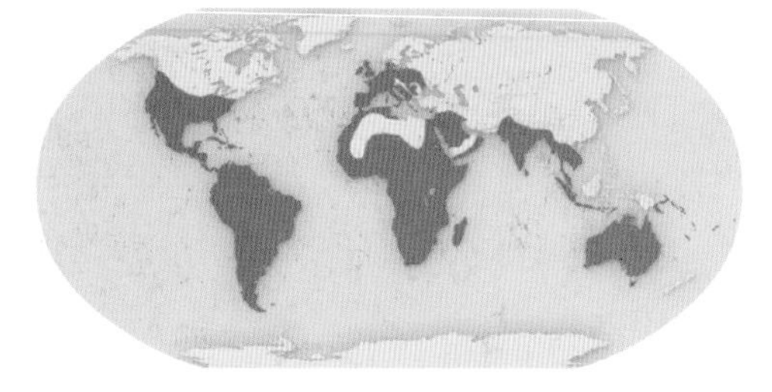

HABITATS AND DISTRIBUTION

The breeding and wintering ranges of most birds are determined by specific habitat requirements. Many species can live and survive in only one type of habitat; some can thrive in a range of different habitats. Those birds whose needs can be met by a number of habitat types are generally more widely distributed throughout the world's zones than birds that are confined to one kind of habitat. Factors such as climate change and human intervention can affect the distribution and range of bird species. The accumulation of pesticides in aquatic food chains, for example, has reduced the distribution of many fish-eating species. Expanding food supplies, on the other hand, have allowed many species of gulls in Europe and North America to extend their breeding ranges farther southward.

↑ **Ospreys** nest in inland lakes and waterways and along coasts in almost all parts of the world. They are extremely adaptable in their nesting habits.

← **Peregrine Falcons** nest everywhere except in Antarctica. Here, one attends its brood of downy chicks in Tasmania, Australia.

Peregrine Falcon distribution

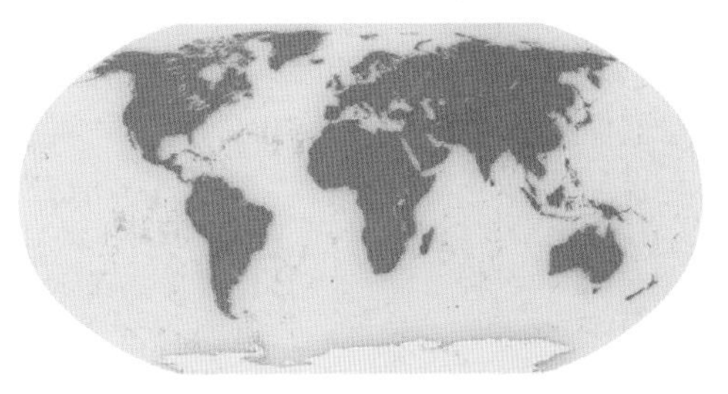

Birds with limited distribution

Some birds are limited in their range to a few mountaintops or tiny islands. Factors that limit the distribution of bird species include evolutionary history, specialized traits and physical barriers. The particular geographic region and specific habitat in which a species evolved can affect its eventual distribution. Many species have not been able to adapt to areas and habitats beyond this "center of origin;" others have managed to colonize quite different environments. Competition for food and territory from other birds, especially closely related ones, can limit a species' potential to expand. Other impediments to a wider distribution include climatic conditions and physical barriers such as mountains, rivers and oceans. The shrinking or destruction of habitats, often as a result of human exploitation or climate change, can reduce the distribution of bird species. The expansion of deserts, or desertification, is a major problem in Africa and has caused a number of species to contract their distribution there. Local extinctions account for the limited ranges of many species and make those species even more vulnerable to further contraction, and even complete extinction. Natural disasters, climate change and human encroachment have a serious impact on species with limited distributions.

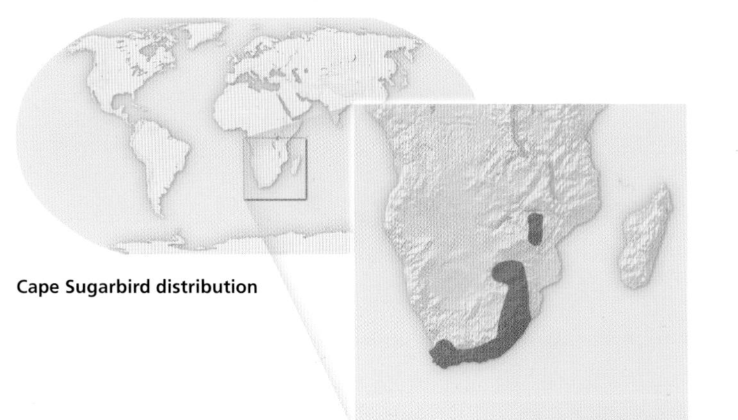

Cape Sugarbird distribution

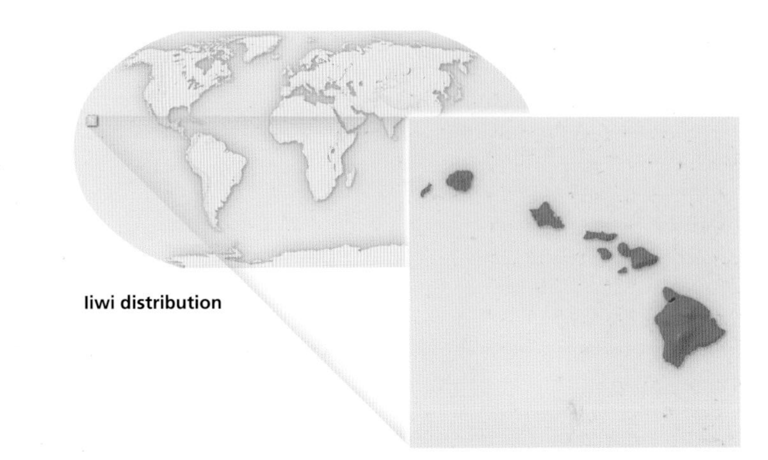

Iiwi distribution

→ **The brightly colored Iiwi** occurs on Hawaiian islands. At least 11 species in its subfamily, which has played an important role in Hawaiian culture, are now extinct.

↖ **Cape Sugarbirds** are restricted to areas of South Africa where Cape Fynbo plants flourish.

← **Galàpagos Hawks** occur on most of the Galàpagos Islands. Here, the hawks are on Isabela Island.

RANGES AND POPULATION SIZES

Species with restricted ranges may not be vulnerable if they have large population sizes. Birds such as the Jamaican Woodpecker, the Puerto Rican Tody or the Amakihi, which is endemic to Hawaii's Big Island, are abundant within their narrow ranges and are not endangered. However, when populations within limited ranges begin to decline, there is cause for concern. This is occurring with many parrot species that exist only on islands in the Lesser Antilles, and with many of Hawaii's endemic birds. Despite intensive management efforts, the population of the Poo-uli, which existed only in a small area on Maui, declined drastically and it recently became extinct. Intensive management has, however, rescued New Zealand's Chatham Island Black Robin from extinction. The extensive cutting of old-growth forests in the northwestern United States has reduced both the range and number of Northern Spotted Owls, which have proved unable to adapt to any other environment.

What is habitat?

Large major habitat types are called biomes. Biomes are generally identified with their dominant vegetation types. Widely known biomes include grasslands, tropical rain forests, coniferous forests, deserts and tundras. Less familiar are chaparrals, savannas, paramos, pampas and scrub deserts. The species density, composition, height and dispersion pattern of the dominant vegetation within a biome type differ with geographic location. Most biomes are complex mixtures of different habitats. Forests may contain lakes, marshes or streams, and can be mountainous or flat; grasslands have marshes, tree- or shrub-lined streams, with grasses of different heights and densities. The bare sands of most deserts are punctuated by small water holes and vegetated or rocky outcroppings. Some deserts have dunes; others do not. Birds can be generalists or specialists in terms of habitat. Some birds, such as grouse and tinamous, live all year round in restricted habitats. Other species migrate in autumn from breeding grounds in dense coniferous forests to much more sparsely wooded areas. Shorebirds that breed in northern tundra on wet muskeg migrate along coastal beaches to winter in inland lakes in the pampas of Argentina. Some species alternate between habitats throughout the year. Ornithologists normally identify a species' preferred or typical habitat as the one in which it breeds. Many migratory birds, however, live for only three to four months in their breeding territories.

↗ **The tropical lowland forests** of Central and South America, with their abundance of fruits and berries, provide an ideal habitat for Keel-billed Toucans. Their bills and yellow bibs provide good camouflage in their forest surroundings.

→ **White-tailed Ptarmigans** live in tundra on high mountains or in Arctic regions. They burrow into the winter snow to keep themselves warm. In their winter environment, they molt into an all-white plumage; in summer their mottled browns and blacks blend with the background colors.

← **The deserts and dry grasslands** of Africa provide a congenial habitat for Ostriches. These, the largest living birds, are relatively safe from predators. They travel around in small flocks of up to a dozen and sometimes move with herds of ungulates.

↙ **Most birds that live in grasslands** have dull colored plumage that acts as a camouflage against predators. The Social Lapwing, which breeds in central Asia, and winters in large flocks in eastern Africa and the Middle East, relies on its cryptic coloration for protection.

CHANGING HABITATS

Most habitats change, but those in areas with extreme climatic conditions change slowly. Sterile sand deserts rarely change because they are maintained by low rainfall and strong winds. High Arctic tundra remains constant because the environmental conditions are similar from year to year. Both temperate and tropical habitats, however, are vulnerable to disruptions such as fires, windstorms, diseases, rainfall variations or the invasion of introduced species. Regular habitat change is known as "succession." When a habitat undergoes change, it reaches a "climax," which is maintained until the disruption brings about further changes. Primary succession begins on a substrate, such as bare rock, sand or fresh lava, which has never previously supported vegetation. When substrate is exposed, soil begins to form. First lichens and mosses, then small herbs and eventually shrubs and trees, grow in the soil and break down the rock. Secondary succession occurs when an existing plant community is disturbed by fire, storms, disease or human intervention. Then the community gradually reverts to its original climax state. All stages in a succession help determine which species of birds can live in a habitat. The early successional stages provide many birds with the conditions they need for breeding and foraging. Grasslands, which undergo rapid succession into shrublands and forest, are among the most endangered of all habitats. They need continual disturbances, such as fires, to maintain them. Some larks, sparrows and shorebirds that breed in grassland habitats have become endangered or threatened.

Oceans and seas

Seabirds feed exclusively in the world's oceans. They obtain their food by plunge-diving, swimming or soaring on wind deflected up from the waves. The ocean and the air above it is their home, and they come to land only to breed. There are two main types of seabirds: pelagic species roam the open ocean, feeding on small fish, squid and crustaceans; coastal species forage in the mud, rocks or shallow waters around mainland or oceanic islands, where they find fish, crustaceans, mollusks and other small invertebrates. Petrels, albatrosses, shearwaters and fulmars are pelagic; penguins, boobies, gannets, cormorants, puffins, gulls, terns and skimmers are coastal species. Tiny floating organisms called plankton nourish many small marine animals that in turn are eaten by the fish on which seabirds depend. When marine animals die, they sink and provide nourishment for organisms on the ocean floor. Some nutrients for marine animals originate on land, are carried by rivers and streams to the coast, and are moved by ocean currents, tides, waves and upwellings. In tropical regions, the sun heats the upper layers of water, which sit above plankton-poor layers of cold water. However, large-scale upwellings, the result of merging ocean currents or the upward thrust of the ocean floor, bring nutrients to the surface. These upwellings provide the lifeblood for millions of seabirds. The world's major concentrations of breeding seabirds occur near them.

↑ **Wilson's Storm Petrels** feed over the open ocean by pattering erratically along the surface, and picking up small marine animals and fatty carrion.

← **Black-browed Albatrosses** can soar for days over the open ocean on strong winds and wave updrafts. They come to land only to breed.

→ **Brunnich's Guillemot** breeds on ledges of steep cliffs and feeds in surrounding waters. It pursues its prey underwater and is often caught in fishing nets.

OCEAN FISHING

Seabirds are highly versatile in the ways they exploit oceanic resources, and they have evolved a wide range of feeding techniques. Birds such as pelicans, terns and gulls plunge-dive for fish, sometimes from great heights. The height from which a bird dives is dependent on the depth at which its prey is located. Many petrels and storm petrels grasp their prey as they patter along the ocean surface. Albatrosses soar above the water, seizing prey from near the surface. So, too, do some petrels and gulls. However, most gulls and petrels, along with storm petrels and skuas, swim in the water and dip down to pick up fish, squid and other prey that is just below the surface. Penguins, boobies, cormorants, auks, puffins and some pelicans dive under the water and swim in pursuit of their prey. Emperor Penguins spend up to three-quarters of their life in the sea, coming ashore only to breed and molt.

Seashores and estuaries

The transition from land to the open ocean occurs along seashores and in estuaries. These are among the most productive biomes in the world—they continually receive nutrients from rivers and streams, and the tides wash away contaminants and wastes. Seashores and estuaries are varied habitats and include rocky bays, extensive grasslands, called saltmarshes, and mangrove swamps that fringe the land. All these places teem with an abundance of marine and bird life. The ribbon of bays and estuaries that border every continent and all oceanic islands serves as the nursery for many species of fish and shellfish, and a foraging ground for a range of birds. As well, a rich invertebrate community lives just below the sand. Many groups of birds, including seabirds, wading birds and some shorebirds, are adapted to breeding along coastlines or the shores of oceanic islands. Cormorants, frigatebirds, boobies, gannets, gulls, terns, herons, egrets, puffins and some auks are just some of the seabirds that breed there. Massive colonies of thousands, or even millions, of seabirds nest in mixed-species colonies on cliffs overlooking coastal upwellings. Many birds that breed in northern inland forests and marshes migrate to the shores and estuaries of Africa, Central and South America and southern Asia, and spend the winter there. Shorebirds that breed all across the tundra also migrate to coastal and estuarine wintering grounds.

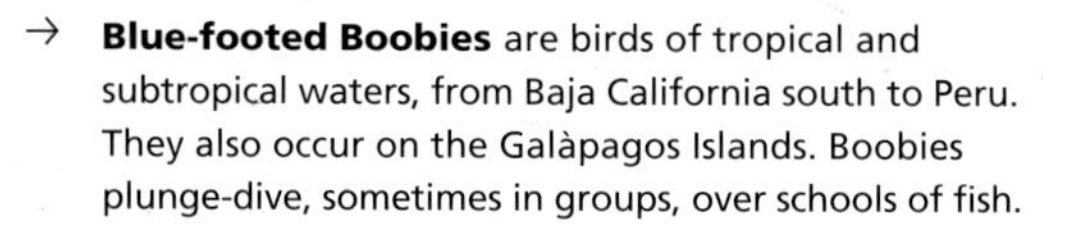

→ **Blue-footed Boobies** are birds of tropical and subtropical waters, from Baja California south to Peru. They also occur on the Galàpagos Islands. Boobies plunge-dive, sometimes in groups, over schools of fish.

↑ **Snowy Egrets** forage energetically along North American lakes, marshes and coastlines. They move rapidly, flapping their wings and often startling fish into coming to the surface. They nest in large colonies, generally along coastlines.

→ **Double-crested Cormorants** and Common Murres nest in massive mixed-species colonies in which each pair has its own ledge or territory. Most colonies are along rocky shorelines or on offshore islands in areas where upwellings bring an abundance of fish.

RECOVERY AND CHANGE

Seashore environments constantly change as tides, winds and storms influence their extent, form and vegetation. Barrier beaches, saltmarshes, mangrove swamps and rocky headlands serve to buffer the land from flood tides, storm gales and hurricanes. Coastal vegetation has adapted to withstand the assault of severe storms, or to recover quickly from any damage they inflict. But recovery is impeded in habitats where artificial barriers have eliminated or degraded sandy beaches or mangroves. While several species of birds nest along seashores, others make their nests in trees or in upland habitats that are less affected by storms. Some shorebirds, terns, rails and sparrows that nest on the ground or in low vegetation are vulnerable to flood tides and changes in the shoreline. Most shorebird and tern species that nest on beaches are adapted to nest on open sand with little vegetation.

Freshwater habitats

Lakes, ponds, marshes, rivers and streams are freshwater habitats for a wide range of bird species. Lakes provide abundant habitat opportunities, although shallower lakes are much richer in plant life and bird diversity than large, deep ones. Marshes, combinations of open water with emergent vegetation, are home to a diversity of birds, which can nest on the shore, on the water, on small islands or in the emergent plant life. Small lakes or marshes in tundra regions have lower species diversity than those in more temperate climes. The birds adapted to freshwater habitats usually have larger than average oil (uropygial) glands, strong, short legs with webbed toes, and dense, oily plumage. Freshwater birds include some ducks, rails, shorebirds, gulls, terns, herons, egrets, wrens, blackbirds and grebes. Herons and sandpipers feed on shores or in shallow water. Other birds swim on the surface and dip or dive under the water to capture insects, other invertebrates and small fish. Diving birds, such as divers and grebes, have sleek, flattened bodies and legs toward the rear of their bodies for smooth diving. Dippers, small songbirds that plunge into rapidly flowing streams, scramble along the bottom in search of food; they can dive for relatively long periods because they lower their heart rate by between 55 and 70 percent and thus increase the amount of oxygen in their blood.

← **Sedge Warblers breed** in northern Europe and Britain and migrate to sub-Saharan Africa for the winter. They build elaborate nests among the reeds in marshes, attaching the base of their nest to dead reeds. Sometimes they nest in low, dense vegetation in wet areas or even in gravel pits.

↓ **The Lotus Bird of Australia** belongs to the family of waders known as jacanas. It lives on still inland waters. It has exceptionally long toes for walking on flimsy lily pads and other thin floating vegetation. Lotus Birds build simple nests on water plants and forage, often in small groups, for a variety of food including fish, insects, mollusks and some seeds.

↑ **An Eared Grebe** in full breeding plumage incubates its eggs on a floating nest that may extend for a foot (30 cm) or more beneath the surface. When it leaves the nest, it covers the eggs with vegetation to hide them from predators. Its bill is open in a scolding display, possibly directed toward the photographer or an intruding neighbor.

SHRINKING MARSHES

North America's prairie region once had extensive freshwater pothole lakes and marshes. They were created some 10,000 years ago by retreating glaciers and were hundreds of acres in extent. Over time, as a result of natural forces or human intervention, the lakes have shrunk. This has caused problems for many resident species: it has resulted in reduced breeding habitat, more territorial competition, fewer food resources and increased threat from predators. Marsh-nesting species evolved in the absence of land-based predators and people. Many birds adapted to the solitude of these marshes, and are unusually sensitive to disturbance by humans. Because they were never frightened away from their nests by predators or people, the young of Franklin's Gulls, for example, did not learn to recognize their parents as early as some ground-nesting relatives. Now, when alarmed, the young scatter and are often unable to find their way back to their nests. Black Terns in North American marshes and Black-headed Gulls in European marshes are also vulnerable to human encroachment and their reproductive rates have decreased as a result.

Forests and woodlands

Forests and woodlands grow wherever there is sufficient water and a suitable climate. Forests are dense stands of trees; woodlands are usually smaller, less dense and more open. Forests grade into other habitats: they provide small stands of trees at the edge of the tundra; they grow along rivers and streams in grasslands and other drier habitats; and they occur in patches in the African savannas. The most expansive forests are evergreen coniferous or broad-leaved deciduous. Northern Hemisphere coniferous forests are called boreal forests or taiga. Their dense shade results in a poorly developed understory that provides good shelter for grouse. Deciduous forests occur in eastern North America, Europe, Asia, northern Venezuela and parts of New Zealand. In many drier places, such as parts of Australia and Africa, there are abundant open thorn forests and scrub habitats, while in the humid tropics rain forest is the natural vegetation. Forests and woodlands generally have high numbers of birds because the complex structure of a forest provides abundant opportunities for nesting, courting and feeding.

BIRDS OF BOREAL FORESTS

Northern Hemisphere boreal forests provide nesting habitats for a range of resident and migrant birds, and their spruce, fir and pine species are a rich source of cones for food. In summer, these forests support bird populations of between 150 and 300 males, of many different species, per 100 acres (40 hectares). Since many species are migratory, these populations drop in winter to only about 50 or fewer per 100 acres. But at any time of year populations can change rapidly. Coniferous forests are subject to influxes of bark beetles and other defoliating insects, which provide sudden pulses of abundant food for insect-eating birds. If birds deplete their sources of this food they will travel elsewhere in massive irruptive movements. Many birds are specially adapted for living under closed canopies. They have short, rounded wings for maneuverability in thick vegetation; melodious songs for attracting mates in places with low visibility; and feet that enable them to walk up tree trunks. Some birds in the boreal forest also have specialized bills for feeding on pine and other conifer cones. The bill of a crossbill is adapted to extract conifer seeds, its sole food, from cones; the multipurpose bill of a jay can handle many food types. In years when conifer cones are plentiful, crossbills remain in boreal regions, but in years of scarcity, they must move to more southerly regions. The flexible jays simply switch to other foods.

↗ **Because the Galah cockatoo is so abundant** in the savanna woodlands of Australia, its beauty is seldom appreciated. Galahs are highly social parrots that travel only in groups. They rest in trees during the hot midday hours.

→ **Mistle Thrushes feed** mainly on the ground in deciduous forests and in gardens around homes. They also feed in trees on ripening berries.

← **Northern Hawk Owls** often survey the forest from the tops of conifers, searching for mammals or other birds to prey on. These normally resident birds only rarely winter far south of their breeding range.

↞ **The morning chorus of the Laughing Kookaburra** is a common sound in eastern Australia. This large bird's disruptive pattern is an effective camouflage in woodlands and open forests.

Jungles and rain forests

Of all terrestrial habitats, tropical rain forests contain the richest diversity of plant and animal life. Rain forests occupy low-lying areas near the equator, where rainfall exceeds 80 inches (203 cm) per year. They occur in the Amazon and Orinoco basins in South America, in the Central American Isthmus, in central and western Africa, in Madagascar, in Southeast Asia and in the Australasian region. The many different levels of trees and shrubs within a rain forest, and the abundant vines and dense ground vegetation, support a wide spectrum of bird species, and each major rain forest has its own unique combination. American rain forests are rich in hermit hummingbirds, tanagers, antbirds, manakins and tyrant flycatchers; sunbirds and warblers are prominent in African rain forests; broadbills, leafbirds, sunbirds and flowerpeckers feature in Southeast Asia; and in the Australian and New Guinea tropics, bower birds and birds-of-paradise are among the many distinctive species. Some rain-forest species, including tinamous and antpittas in America and Chowchillas in Australia, are dull-colored to blend in with the forest floor; others, such as the brilliant orange Cock-of-the-rock in South America, are brightly colored for display purposes.

A THREATENED HABITAT

Jungles and rain forests are the most endangered of all habitats. They are disappearing from Central and South America and from Asia and New Guinea. Nearly half of Earth's rain forests have been destroyed or degraded in the last 50 years for fuel or agricultural development. Many researchers predict that half the remaining rain forest will disappear by 2022 unless drastic action is taken. Not only are areas being clear-cut; forest patches are being isolated, leaving habitat islands where large birds can no longer survive. Army ant swarms disappear; soon after, the birds that feed on these swarms also disappear. Birds that migrate seasonally up and down mountains need rain-forest corridors that are at least 1000 feet (300 m) wide to move between habitat patches. Bird conservation requires an approach that keeps these patches connected, and that ensures the patches are large enough to preserve the populations of birds that nest in them.

← **Andean Cock-of-the-rock males** stand on their lek in the Peruvian cloud forests at elevations of almost 5000 feet (1500 m). Males are promiscuous and display to attract females. The Jivaro Indians of South America copy their display in a sensual ceremonial dance.

→ **Blue and Yellow Macaws** in flight make an eye-catching spectacle in the tropical rain forests of South America. This species is social, and they usually travel in small flocks or in family groups. Pairs typically mate for life.

← **The Southern Cassowary** is a rain-forest dweller. Its predominantly drab plumage blends with the dark shadows of the forest, though its bright blue head and red wattles stand out like flowers against the green vegetation.

↞ **The Blue-crowned Motmot**, the most widespread of the 10 motmot species, has distinctive racquet-tipped central tail feathers. It sometimes preys on butterflies in Central and South American rain forests.

Polar regions

Long winter snows, and cold temperatures and winds typify polar regions. Only the upper layers of soil ever thaw. Below is permafrost—permanently frozen land. Polar regions extend north from the tree line in the Northern Hemisphere, and south from South America, Africa and Australia in the Southern Hemisphere. In winter, daylight is brief; in summer it is almost continuous. In the Arctic, grasses, sedges, mosses, lichens and dwarf willows grow in hummocky places low to the ground; in Antarctica, only a few mosses and liverworts can survive. Despite the rigors of a short growing season, sparse vegetation, a lack of trees and extreme temperatures, some birds remain throughout the year. Ptarmigans, Gyrfalcons and Snowy Owls are permanent residents in the Arctic, though their survival strategies differ. Ptarmigans stay in the same places all year, eking out a living on berries and other vegetation under the snow. Snowy Owls travel nomadically during winter in search of lemmings; if these become scarce, they move to temperate regions to find mice.

ARCTIC AND ALPINE TUNDRA

The climatic conditions that produce the polar tundra can occur high up on mountain peaks. At higher elevations, boreal forests become shorter as they approach the tree line. The alpine tundra in North and South America features rugged, well-drained soil interspersed with meadows that are lush in early summer. Meadow lakes and streams provide places where a few willows can grow taller than the mainly ground-hugging vegetation. In both alpine tundra and Arctic tundra there are freezing temperatures all year, low vegetation and a short growing season. Alpine tundra usually has high winds, intense sunlight, a lack of permafrost, fewer extremes of light and darkness and reduced oxygen. Bird diversity in alpine tundra is low—fewer than 50 males, of all species, per 100 acres (40 hectares). The American Pipit, the White-tailed Ptarmigan and several species of rosy finches occur in both the alpine tundra in North America and in Arctic tundra. No predatory species nest regularly in alpine tundra, whereas owls, Gyrfalcons and jaegers nest and breed in Arctic tundra.

← **An Adelie Penguin** stands 2.5 feet (75 cm) tall. It breeds only on the shores of Antarctica, where it makes a nest of small rocks for its single egg. These birds nest in large, noisy colonies and usually travel in groups. They take to the ocean to seek food and protection from the extreme cold on land. Here, a lone Adelie Penguin walks, in typically upright fashion, across an iceberg.

↟ **Snowy Owls** are adapted to the Arctic, and are circumpolar in distribution. They grow 2 feet (60 cm) tall, but their eyes are almost as large human eyes.

↑ **The dark plumage of the Northern Raven**, one of the largest of the raven family, absorbs heat. This helps it to withstand the harsh climate in which it lives all year round.

Grasslands and moorlands

Vast tracts of unbroken, swaying grasses dominate the interior of many continents. Grasslands—called prairies in North America, pampas in South America, moorlands in Europe and steppes in Asia—occur where rainfall is moderate and terrain is flat. African grasslands, called savannas, have scattered trees. The density and height of grasses depend upon the amount and seasonality of rainfall. Where rainfall is reduced, grasses are shorter and trees occur only near streams. The diversity of grass species in these ecosystems provides a range of seeds for seed-eating birds. Grassland birds have adapted to the open and exposed habitat by hiding nests on the ground or in low vegetation. This protects them and their nests from predators and from exposure to the sun and winds. A few species even nest underground, in burrows. Some males display aerially, soaring upward to attract females or announce territories. Birds that live in grasslands are typically streaked and colored mottled brown. Because of the relative scarcity of nesting and feeding places, breeding bird density is generally low. Often it is fewer than 100 males per 100 acres (40 hectares).

↙ **Burrowing Owls** nest in underground burrows. With their long legs, they move easily in the thick grasses.

↓ **Bobolinks** occur only in grasslands. Typically, they sing in flight, but may also call from the tops of swaying grasses.

↘ **Grassland fires** flush out insects, which are readily devoured by birds such as these Yellow-billed and Cattle egrets, as well as other species.

→ **African White-backed Vultures** are gregarious. They feed on soft tissue, such as the muscles and intestines of dead animals. Several hundred may descend together to feed on a carcass.

EXPANDING THREATS

Most grassland birds, in all parts of the world, are threatened by the expansion of agriculture and grazing. Rich, fertile land can support agriculture where there is sufficient water; cattle and sheep grazing requires less water. Grassland birds are declining more steadily, and over a much wider geographical area, than any other habitat group. In North and South America, shrikes, meadowlarks and grassland sparrows are decreasing at alarming rates, and some grassland species are facing extinction at both local and regional levels. Until fairly recently, farmers cultivated or grazed only part of their land; the remaining hedgerows and grasslands were sufficient to maintain communities of nesting birds. Now the hedgerows and grassy edges have disappeared, and hayfields are harvested earlier, often during the nesting season, when eggs and chicks are most vulnerable. The use of pesticides in grasslands has destroyed the insects on which many grassland birds feed. Some North American species, such as the Bobolink and the Dickcissel, migrate in the winter to South American rice-growing areas, which are also subject to heavy chemical spraying.

High mountain regions

Harsh climatic conditions typify most high mountain regions. Different elevations have distinctive populations of plants and animals. In many areas, low temperature and moisture levels combine to create vegetation-free zones near the tops of mountains. The tree line—between forest and tundra—provides visual evidence of altered environmental conditions. Above the tree line the conditions are harsher, with higher winds and greater variations between day and night temperatures. The tops of some mountains, ridges and slopes have alpine tundra habitat, with only low vegetation, windswept rocks and marshy pools. Crevices in rocks and tundra meadows provide opportunities for nesting and foraging during the warm summers, but are inhospitable during winter. Marked seasonal variations force birds either to adapt to the changing conditions or leave. Many birds simply move up and down mountains as seasons change. By moving down into grassland or forest, they can find cover and food during the winter. These vertical movements allow some montane birds to remain resident in a region. Other birds adapt, either by migrating to warmer climes or by adopting nomadic lifestyles. Nomadic species move when seasonal conditions do not provide enough food.

↓ **Eurasian Nutcrackers** fight in winter over limited food supplies. Sometimes, these birds migrate irruptively to areas where food is more readily available.

→ **The Blue Grouse**, shown here, and closely related Spruce Grouse are adapted to survive winters in harsh North American montane conditions. Evergreen pines afford them protection from the wind and cold.

→ **Blood Pheasants** live in coniferous forests above 10,000 feet (3000 m), and often at the snow line, in the mountains of China, Tibet and Nepal. Their striped coloration blends with the undergrowth and with the tree trunks and branches, where they fly to avoid predators.

← **As they search for carrion**, Andean Condors soar high on updrafts above mountain peaks and valleys. They nest on cliff ledges, and sometimes they glide down to the coastlines between Colombia and Chile.

MALES ON DISPLAY

In spring and summer, Blue Grouse and other grouse species nest near the tree line in Northern Hemisphere coniferous forests, and retreat down the mountain in winter. Male grouse have elaborate displays or bright coloration to attract females. They adapt to the dark conditions of montane forests by making loud booming sounds that travel well within the forest. Once the sounds have attracted distant females into closer proximity, the males entice them further by spreading their tails in a wide arc. In Asia a wide range of pheasants fills a similar niche in the montane regions. Many of these pheasants have elongated and elaborately colored tails, and are among the showiest of birds. The males display these ornate tails, which they can vault or arch into inverted V-shapes, in territorial contests and to attract females. With their tail folded, they can creep through the forest, blending in with the leaves and small herbs. Three of the most beautiful pheasants—the Golden Pheasant, Silver Pheasant and the Lady Amherst Pheasant—live in the mountains of China. The magnificent male Reeve's Pheasant has a tail so long that the bird measures 5 feet (1.5 m) from the tip of its beak to the end of its tail. Reeve's Pheasants are elusive; if surprised, they can retreat soundlessly into the montane forest.

Desert regions

Deserts occur wherever there is very low rainfall. In some deserts, rain never falls. Low humidity, strong winds, bright sunny days, extreme seasonal variations and abrupt drops in temperature at night are also characteristic of deserts. The daily temperature range in deserts is greater than in any other habitat. Where there are winter rains, water drains into slight depressions in the sand. The only plants that can survive the rigors of desert conditions are those, such as cacti in the New World and euphorbs in the Old World, that can withstand desiccation or that can store water. The appearance and form of deserts vary greatly. Some are characterized by sweeping dunes and a complete lack of vegetation; in others, tall branching cacti or thorny bushes grow abundantly. The harshness of the conditions and the availability of food and shelter determine the species and numbers of birds in desert environments. Where vegetation is sparse and water scarce, bird diversity is low and populations are small. In deserts where there is a greater variety of plants, the presence of insects and reptiles can sustain a range of bird species. As well, the vegetation provides nesting opportunities and hiding places.

↑ **Sandgrouse** are cryptically colored and travel in large flocks. Here, Spotted Sandgrouse stop to drink at a waterhole in the Okavango Delta, Botswana.

DESERT ADAPTATIONS

Desert birds have adapted to the scarcity, or lack, of surface water and the extremes of temperature. They cope with the heat by remaining inactive during the hottest part of the day, hiding behind or under vegetation and, in rare cases, taking refuge in the burrows of desert mammals. Some desert birds do not drink; they obtain all their fluid from their food. Other birds get their water from water holes, dew, raindrops or plants. Nearly half the birds in Australian deserts never drink water. Others go to waterholes, but these visits expose them to predators, who lay in wait. The cryptic coloration of some species allows them to sneak in unobserved at dawn or dusk and evade would-be predators. Other species arrive in large flocks, relying on flock members to warn them of impending danger. The tight flying pattern of these birds, too, often confuses potential predators.

← **Desert Wheatears** live in deserts with sparse vegetation. They often use high vantage points to search for predators.

↑ **Cactus Wrens** nest among the spines of cholla cactus or in thorny shrubs of North American deserts. When they cannot find insects, they resort to eating cactus fruit.

← **Crab Plovers** breed in small colonies along the desert shores of the Indian Ocean. They lay a single, large egg at the end of a tunnel which they dig in the sand.

Urban regions

Nearly half of the world's people now live in cities, and most of the rest live in towns and villages. Towns and cities provide a wide range of bird habitats, and many species have adapted to them. Most urban areas have parks, tree-lined streets, lawns, small gardens, window boxes and patio gardens. Shrubs, trees, poles, houses, window ledges and rooftops are all potential nesting sites. To live in urban environments, birds need food sources, safe nesting places and protection from human interference. In urban settings they can forage for insects on trees and shrubs, for berries on street and park plants, for seeds in lawns and nature strips, and for a range of foods in garbage. Many species readily accept food from humans. Starlings, sparrows and pigeons are familiar city birds. So, too, are numerous songbirds and some species of hawks, vultures, crows, jackdaws, parrots, herons and egrets. Some birds have learned that by roosting in city trees they are relatively safe from predators. Species that roost in towns include parakeets in India, Cattle Egrets in the southern United States, storks in Africa and Rock Pigeons and Eurasian Starlings almost everywhere. Asian Open-billed Storks frequently nest in the protected grounds of monasteries. Lakes in many cities provide safe wintering grounds. Large numbers of ducks live in ponds in downtown Tokyo, Japan, where the zoo provides a safe breeding site for colonies of Great Cormorants. Cockatoos and ibises are common sights in the parks of Sydney, Australia.

↑ **Common Starlings** are among the most urbanized bird species. They frequently exploit urban structures for communal roosting.

← **Eurasian Kestrels** have readily adapted to living in towns and cities, where they forage among the shade trees and along roads.

→ **In the wild, Canada Geese** live in tundra habitats. These four geese are on a frozen lagoon in Chicago, in the American midwest.

NESTING OPPORTUNITIES

City life is proving increasingly congenial to both colonial- and solitary-nesting birds. Peregrine Falcons, a solitary species, use window ledges, which can resemble cliff ledges, as nest sites. For many years a pair of Red-tailed Hawks has raised its young on a Manhattan high-rise. Monk Parakeets attach their communal stick nests to telephone poles and wires in cities in North and South America. Doves everywhere nest on window ledges, in flower boxes and over front doors. In San Juan, Puerto Rico, Rock Pigeons nest in holes in churches and other buildings. Herons and egrets form dense colonies in the centers of many towns in Mexico and India; during the day they fly out of town to feed in marshes and along rivers. Vultures roost in Indian towns, where they feed on carrion on roadways. Ibises in Japan, Indonesia and Egypt roost and nest in towns and forage in nearby rice fields.

Farms and gardens

For millennia humans have cultivated farms and gardens, and birds have long adapted to the range of habitats they provide. Domestic gardens, with their variety of trees and shrubs, are a haven for many nectar-eaters. Farms vary greatly in size, function and agricultural methods. Some are monocultures that grow only one kind of crop; others grow a range of crops and raise or graze animals as various as sheep, cattle, horses, fish and shellfish. As with other habitats, the diversity and placement of vegetation and the availability of water determine the species and population sizes of birds that can live there. Monoculture farms generally provide fewer nesting and foraging opportunities than multi-functional farms. Small woodlots, hedgerows and the buildings associated with farms afford shelter and protection from predators and inclement weather. Among the groups that have adapted to farmlands and gardens are some species of shorebirds, gulls, hawks and sparrows, as well as thrushes and other songbirds. Ducks, geese, cranes, sparrows and blackbirds forage on farms during migratory journeys. Cowbirds, starlings, grackles, Cattle Egrets and some flycatchers are common around livestock. Wet agricultural environments, such as rice paddies, are year-round foraging habitats for herons, egrets, ibises, storks, ducks and cranes. In winter, Bobolinks and Dickcissels feed on the rice.

↑ **Corvids, the diverse bird group** that comprises crows, jays and magpies, are highly adaptable. Many species, such as this European Magpie, are commonly seen on farms and in gardens, as well as in towns and villages, in North America, Europe and throughout Asia.

→ **A Winter Wren** feeds its hungry young in the globular nest that it has built high up in a farm building's cavity. These birds seek out similar crevices and nest boxes, which are usually safe from predators, throughout northern Europe and North America.

BIRDS AND FARMING TRENDS

The relatively recent, and rapid, transition from predominantly small-scale to large-scale farming in many parts of the world has reduced the habitat diversity available to birds. For thousands of years, people farmed only small plots. When the soil was depleted, they moved on to other areas or cultivated new fields. Abandoned farms gradually reverted to their former native condition. As a result, rural areas were a mosaic of habitats in various stages of succession, from recently plowed fields to woodlots or secondary forest. The existence of farm animals, along with crops, further enhanced the range of habitats in agricultural regions, making it possible for a variety of birds to coexist in a limited geographical area. With the consolidation of farms into bigger parcels, and the trend toward greater crop specialization, habitat opportunities have declined. Limited or monoculture crop farming has brought an increase in plant and insect diseases and made necessary a greater use of pesticides. The result has been a disturbing decline in the kinds and numbers of birds on farms.

↑ **Tractors plowing or harrowing fields** provide opportunities for birds to catch insects, earthworms and grubs that are exposed in the newly turned soil. Here, a dense flock of Black-headed Gulls follows a tractor in search of the juicy pickings that it brings to the surface.

→ **Swallows** are more dependent on humans than most birds. Many swallow species nest mainly on or in structures such as barns, bridges, churches and abandoned buildings. Here, a Barn Swallow leaves a stable on its way to forage.

Bird-rich habitats

Birds are not evenly distributed. Some species range widely across the world; others are endemic to particular regions or habitats. Resources, such as food, foraging habitat and nest sites, determine how many species can coexist in one region. Generally the number of species, or the "avian density," is high in the tropics and low in polar regions. Species diversity decreases with altitude. For example, a mountain peak in the tropics may support fewer than 50 species; in a more temperate region lower down the same mountain there may be 200 species; while the tropical lowlands at the base of the mountain could be home to more than 500 species. There are similar variations between geographic regions. South America has over 3000 species. More than 1500 of these live in Colombia. Greenland, with a land area much larger than Colombia's, supports a mere 56 species. There are more than 1500 species of birds in northeast Africa, between 1000 and 1500 species in southern Asia, but only about 700 species in Australia and New Guinea. Madagascar, northern Africa, Europe and North America north of Mexico each has under 700 species. Islands generally have low species diversity, but a high proportion of endemic species.

↑ **Rhinoceros Hornbills** perch on trees high above the mixed dipterocarp rain forest of Sarawak. Also known as the "Land of the Hornbills," Sarawak is rich in biodiversity; 530 different bird species inhabit its forests.

← **The Straight-billed Hermit** lives in the tropical rain forest of the Amazon, an area with the world's highest bird diversity. It is one of many species of hermits, which are mainly limited to tropical rain forests.

↞ **The Rouget's Rail** is endemic to the Bale Mountains of Ethiopia and often stands on exposed rocks, where it is easily observed. It lives at elevations above 4000 feet (1200 m) and feeds on a range of aquatic insects as well as on snails and seeds.

→ **Birds that share** the same general habitat often occupy different niches within that habitat. This makes for greater species diversity. As shown in this illustration, a variety of insect-eaters can forage and nest in the same European woodland because they each occupy a specific niche in the habitat.

TROPICAL RICHES

Climate stability plays a key role in determining the number of bird species that will live and breed in a region. Biologists believe that the high number and diversity of bird species in tropical places is directly attributable to the relative mildness and predictability of the climate, as well as to the abundance and complexity of the vegetation and the reliability of the food resources. Because tropical climates have been more stable than those in other parts of the world, they have provided greater opportunities for the evolution over time of new species and the creation of new niches. A niche can be defined as the particular role a species plays—the space it occupies and the food it eats—in its environment. In comparison with more northerly regions, tropical environments provide the birds that live and breed there with a wider range of plant and animal food sources and an increased number of breeding habitats.

FOREST NICHES OF INSECT-EATING BIRDS

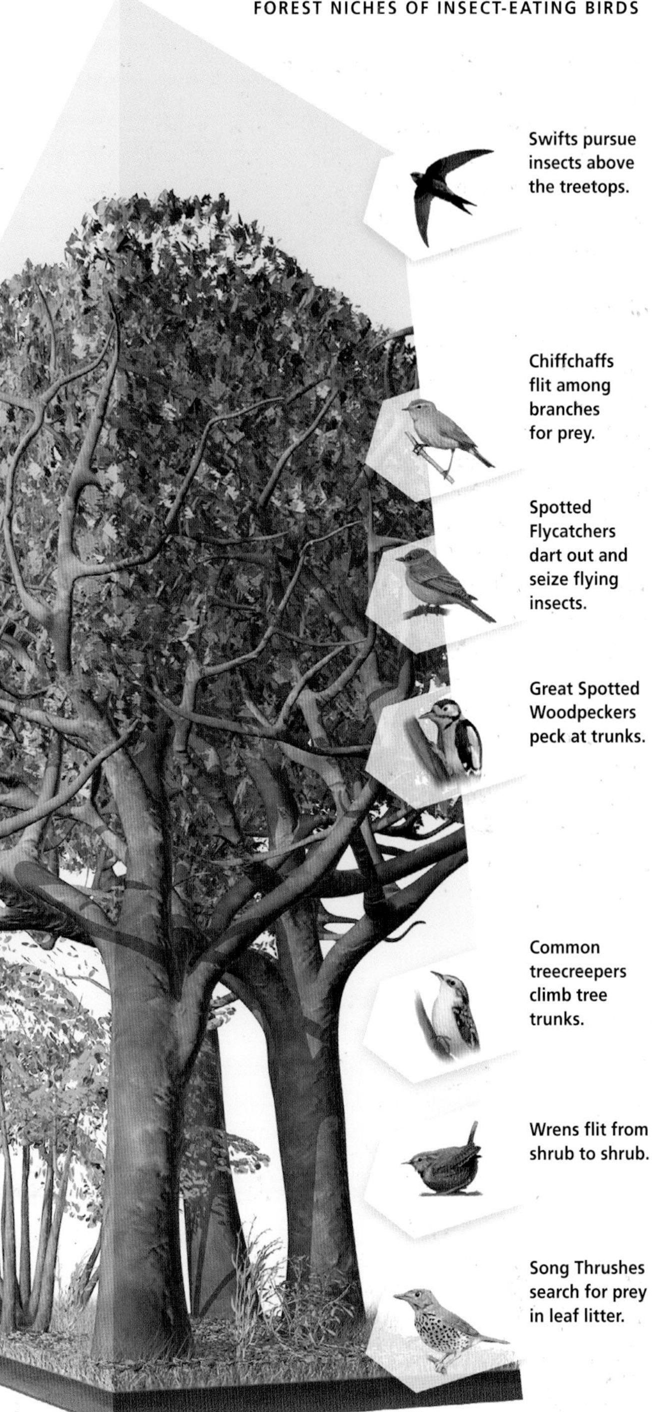

Restricted habitats

Birds that have restricted habitats and ranges often occur only in small numbers and are usually endemic to a specific geographical area. A bird's range can be limited because of naturally imposed restrictions or as a result of human intervention. Island species are often restricted because the birds have never dispersed more widely, or because they were unable to compete when they moved to other environments. Some birds have been forced into island and other small habitat areas when, for example, predators were introduced into their previous, more extensive ranges. Other isolated habitats include mountaintops and forest patches that are surrounded by inhospitable areas. Any barrier that prevents movement can isolate a species. Rivers, oceans and deserts are all potential barriers; so, too, is destroyed, degraded or reduced habitat area. Red-cockaded Woodpeckers in the United States are diminishing in numbers because they are dependent for food and nesting upon dwindling numbers of old-growth pine trees. Piping Plovers in the eastern United States, Hooded Dotterels in Australia and White-fronted Plovers in east Africa are restricted because they can survive only on barren sandy beaches. The Takahe is an example of a species that is extremely limited both in population numbers and habitat. This large, flightless gallinule exists only in two New Zealand South Island mountain ranges and feeds only on tussock grass that has been overgrazed by deer. Fewer than 200 Takahes exist in their natural habitat. Some birds in mainland areas also have restricted ranges. Fewer than 300 Lear's Macaws remain in the semiarid Caatinga habitat of Brazil; they nest on cliffs and fly many miles over inhospitable scrublands to feed on the fruit of only one species of palm tree. These large, spectacular bright blue birds are vulnerable to poaching for the pet trade.

↑ **The Madagascar Paradise Flycatcher** lives only in Madagascar and the nearby small islands of the Comoros group. It perches high up in its forest habitat and swoops down swiftly to catch insects. The male uses its long, white tail feathers to attract females.

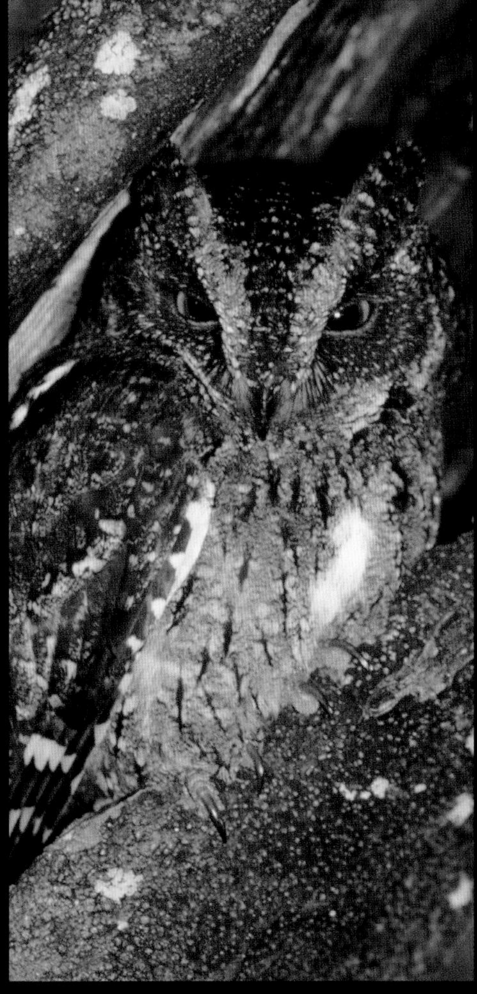

↑ **The small and slender Malagasy Scops Owl** is endemic to Madagascar. Camouflaged in its nest cavity, it sleeps away the day.

↑ **The Swallow-winged Puffbird** is restricted to South American riverine habitats, where it nests in riverside burrows.

← **The highly endangered Imperial Parrot** is the national bird of Dominica, where it is known as Sisserou. It is endemic to the island, but only 60 of this species remain in the wild. A number of other Caribbean Islands also have parrots that are unique to them. These parrots are often poached for the pet trade, and this picture is of a captive bird.

ISLAND PERILS

Island-nesting birds are particularly vulnerable to predators, especially introduced ones. Birds generally do not change their behavior in response to new predators. A species that nests on the ground will continue to do so; a bird that relies on cryptic coloration will not adapt to cope with a newcomer that can locate it by heat. For example, the brown tree snake has eliminated most of Guam's native birds, all of which nest on or near the ground. Unchecked by previous natural predators, the introduced snakes increased rapidly and devoured the eggs and young of all birds they encountered. Many native Hawaiian birds have been displaced by exotic species or have been eaten by cats, mongooses and feral pigs. Some native Hawaiian birds have survived only by nesting on islands that predators have not reached. Cats and other introduced animals have eliminated many of New Zealand's ground- and shrub-nesting birds. Remnant populations of many species, such as the Black Robin, are restricted to offshore islands where predators are strictly controlled. The Ascension Frigatebird has disappeared from Ascension Island and now nests only on offshore rocky islets.

Adaptations for survival

Adaptations for survival

The body parts of birds are adapted to suit their different feeding behaviors and the niches they occupy. Bills, feet and wings all provide clues to birds' lifestyles. While some birds spend their entire lives in one habitat, others travel impressive distances each year to survive.

Threats and responses

Birds live in a dangerous world. In order to survive, they have to avoid predators; adapt to the incursions and depredations of humans; cope with inclement weather and climatic irregularities; and defend their territories, nests, young and mates against competitors. At the same time, they need to find enough food for themselves and their young. Frequently they must defend the food they have found from theft by other birds. Some species are specialist food stealers that have highly developed theft, or "piracy," skills. Birds have evolved a range of mechanisms to help them survive in their environment. These include group movement and migration, aggressive defense strategies, and cryptic coloration and evasive behavior. Some birds feed, breed, roost and migrate in groups, and depend upon their groups for defense and early warning of approaching predators. Others rely for protection on camouflage, or on their capacity to flee swiftly. Some birds are able to vary the timing of their breeding in a way that allows them to avoid predators and take advantage of optimal foraging opportunities. Budgerigars in Australia and Red-billed Queleas in Africa are nomadic species that move in huge flocks, of thousands or millions, waiting for optimal foraging conditions before they can breed. Rains, which bring the growth of lush grasses, trigger their breeding behavior.

→ **The chicks of the Great White Heron**, a color phase of the Great Blue Heron that is confined to Florida, hatch asynchronously, allowing the largest chicks to have best access to food. The chicks are able to climb about the nest within a few days of hatching, and the older chicks sometimes kick the younger chicks out of the nest to avoid food competition.

↓ **Rock Ptarmigans rely on cryptic coloration** to avoid predators. During summer, males and females are brown to match the vibrant tundra. In the fall they molt into an almost pure white plumage, making it difficult for predators to spot them in the Arctic snow. In winter they feed on whatever seeds and vegetation they can find above the snow. This is a female. It lacks the black line that in the male stretches from the bill to the corner of the eye.

ASYNCHRONOUS HATCHING

Some species have varying clutch sizes. When food abounds, species such as herons and egrets may lay five or six eggs; when it is scarce they lay fewer. Food can be plentiful during egg laying, but scarce during chick rearing. Some species respond to this discrepancy by beginning incubation with the laying of the first egg. As a result, chicks hatch at different times, or "asynchronously," and are soon of greatly differing sizes. In some years, there is enough food for all; in others, only the largest chicks can survive. This adaptive phenomenon ensures that in every year at least some strong, healthy chicks will fledge successfully.

↓ **Gray-headed Gulls attack** an egret in order to steal its fish. Young gulls need time to master this difficult foraging technique.

Diverse bodies

Birds' bodies both determine and reflect their ability to survive in their varied habitats. Both individually and in combination, the different parts of a bird's body are adapted for a particular lifestyle. Food types determine the shapes and sizes of bills; foraging depth in water dictates the length or placement of legs. Bodily features are keys to both behavior and habitat. Streamlined birds with short, strong wings and with legs placed far back on their bodies—such as divers, penguins and cormorants—are excellent swimmers. Albatrosses and petrels have streamlined bodies with long slender wings for gliding. Birds such as Ostriches, rheas, bustards, herons, egrets, ibises and cranes have long legs for walking through water or tall grass. Birds that paddle through water have webbed feet and bills that suit their foraging methods: long thin bills are for catching fish; most serrated ones are for straining; and short rounded ones are shaped for pulling up vegetation. Unusual lifestyles make for unusual shapes and sizes. Sunbirds that feed on flowers with long, curved nectar chambers have long, curved bills that enable them to reach deep into these flowers. In some cases, only one part of the anatomy tells us much about a bird's habits. The clue to a songbird's diet, for example, is to be found in its bill, rather than in its rounded body shape and middle-length legs. In general, thin bills are for catching insects; stout, conical ones are for cracking seeds.

← **Avian feet have evolved** to support the ways in which different birds move and the kinds of places they land on, perch on, walk on and swim through.

Paddling in water

Grasping and picking up

Walking and running

Walking on water plants

SHAPED FOR THE COLD

A number of factors combine to determine the size of a bird's body: its lifestyle, habitat and geographical location. Body shape is usually a function of feeding and nesting habits as well as of strategies for avoiding predators. Climate, too, affects both the size and shape of birds. All warm-blooded vertebrates that live close to polar regions tend to have larger, rounder bodies and shorter appendages than those that live in warmer areas. This is known as Allen's Rule, after the nineteenth-century ornithologist Joel Allen, who made this observation. It is based on the principle that bodies of equal volume can have different surface areas. Large, round bodies with small appendages have smaller surface areas from which to shed heat than do small, elongated bodies. The greater its surface area, the more heat a body will lose. Birds, which are warm-blooded animals, need to maintain their quite high body temperature. Many of the birds of polar regions, such as ptarmigans and Snowy Owls, are fat and round. No non-migratory birds that live in polar regions have long, thin legs or long, thin bills. Similarly, resident birds in the cold alpine tundra, such as grouse, also have bulky, round bodies.

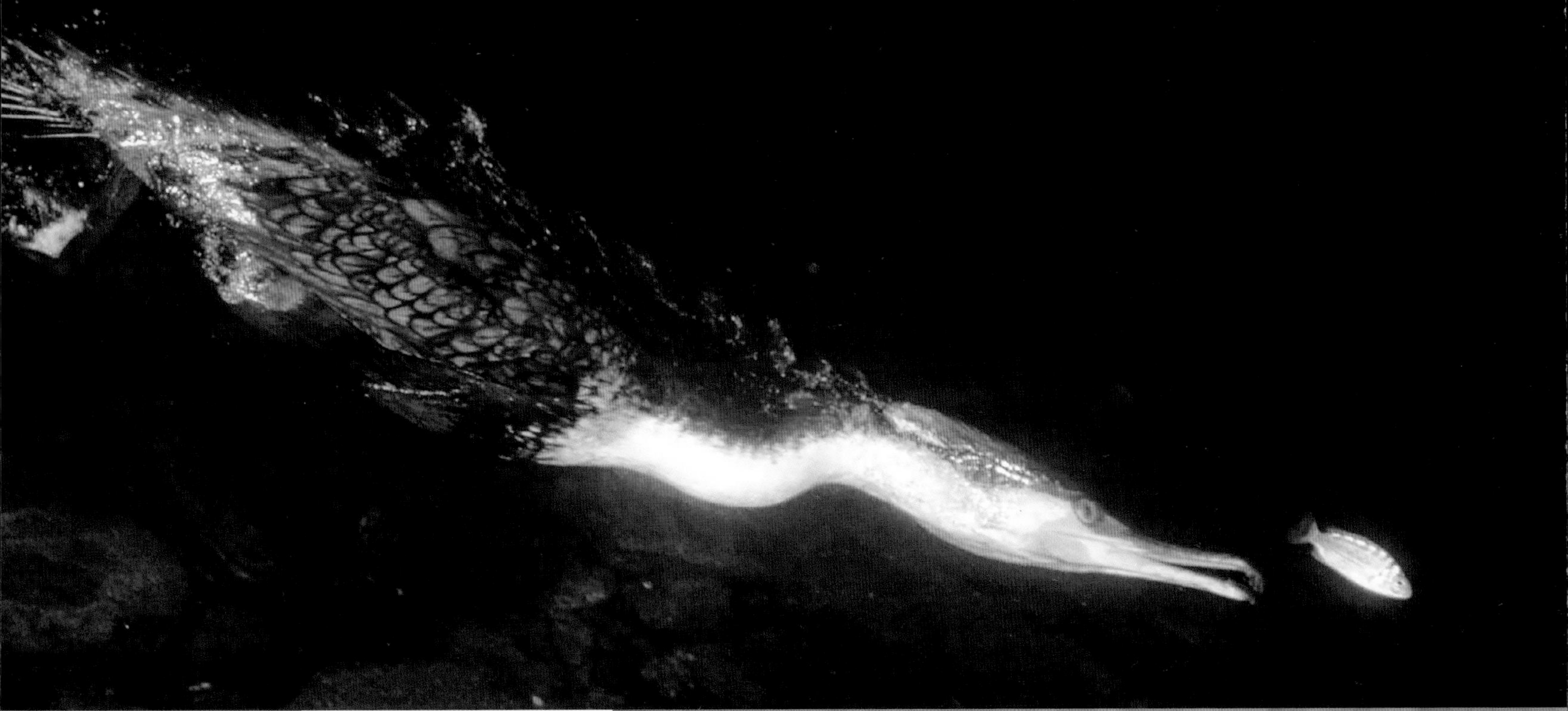

↑ **This Great Cormorant**, its streamlined body adapted for rapid underwater movement, is about to capture a fish.

← **Lammergeiers**, also called Bearded Vultures, are built for soaring. They feed mainly on the large bones of animal carcasses, which they break open on rocks.

→ **A Greater Rhea's long legs** enable it to run swiftly away from a predator or to repel it with a powerful blow.

Feeding strategies

Diet and foraging habitat have been major driving forces in the evolution of birds. A bird's overall shape and size, and the length and form of its bill and legs, are adapted for the kinds of foods it obtains. Birds' diets are highly diverse. Food types include seeds, nuts, fruits, plants, nectar, fish, crustaceans, frogs, toads, snakes, lizards, as well as live birds and mammals and carrion. The size and strength of a bird's bill will depend upon what it needs to crack or open, or what it has to tear apart. Often birds that have similarly shaped bills eat totally different foods. Finches, parrots and hawks all have quite thick, strong bills relative to their head sizes. Finches use their bills to crack open seeds; parrots crack open hard nuts; and hawks tear apart the flesh of birds and mammals. Herons, egrets, ibises, flamingos and some shorebirds have relatively long bills for probing a range of foods. Herons and egrets feed in shallow water on fish and frogs; ibises and some shorebirds forage in the mud for small invertebrates; and flamingos have specialized bills that strain invertebrates from the water. The lower bill of a flamingo is large and troughlike, while the upper bill is thinner, like a lid; the pumping of the throat and the rapid, piston-like movements of the thick tongue suck up currents of water, which the flamingo then filters through serrations in its bill.

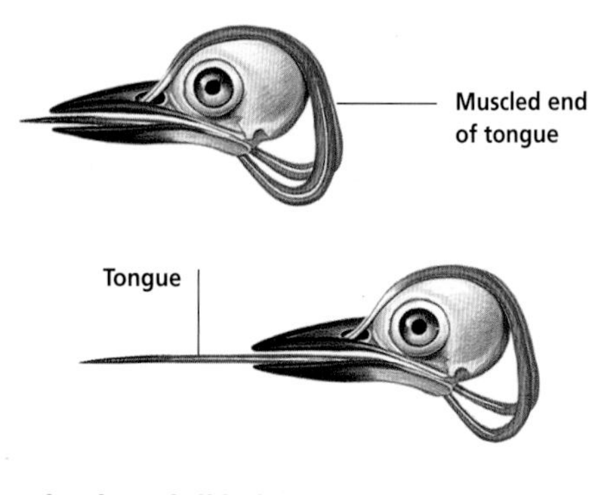

↑ **Woodpeckers drill holes** in trees with their bills. They then extend their long tongues deep into these holes to extract insects. The muscled end of the tongue loops over and behind the skull.

↖ **Song Thrushes** smash the shells of garden snails onto hard surfaces to crack them open and expose the soft tissue within.

↑ **A New Holland Honeyeater**, a bird of Australia, probes the nectar in a flowering Banksia plant.

← **Common Buzzards** are widespread in Europe. Here, one spreads its wings to break its flight before landing and pouncing on its prey.

↞ **Green Woodpeckers** use their strong bills to dig into wood to find insects. They nest in holes in trees.

TONGUE ADAPTATIONS

The way a bird obtains and eats its food will determine the size, form and other special features of its tongue. Woodpeckers use their long tongues, which have backward-projecting barbs at the tip, to pry insects from the crevices of trees; sapsuckers employ their short tongues, which have forward-facing hairs, to draw sap out of holes they have drilled in trees. Bananaquits have forked tongues with fine, brushlike hairs that can absorb nectar and fruit juices. Hummingbirds and sunbirds also have specialized tongues that are adapted for sucking nectar; their long forked tongues curl toward the base to provide a channel, or trough, through which nectar can pass into the mouth without spilling out.

Specialist feeders

The demands imposed by the needs to forage, avoid predators and reproduce can involve a number of specialist adaptations. The form and size of the bill, feet, talons and body of any bird are adapted to serve each of these purposes to varying degrees. Some birds feed on only one type of food; others change diets during the breeding period or from one season to another. Terns are specialists in plunge-diving for fish, and their entire bodies are adapted for this. Their dull coloration makes them less visible to their prey and their streamlined shape allows smooth entry into the water. At the other extreme, many songbirds eat seeds during much of the year, but often supplement their diet with fruits and insects and feed their young insects, which are more energy-giving and digestible than seeds. The conical bill of a Song Sparrow is well suited both to a winter diet of hard seeds and to catching soft-bodied insects in summer. Many shorebirds have long, narrow bills with which they can probe at different depths in the mud and also pick up invertebrates on the surface. Legs and talons are often adapted to achieve the best compromise between movement, stability and food acquisition. Hawks have strong legs and talons for capturing prey and carrying it back to their nests, while the legs and claws of most songbirds do no more than enable them to stand on branches or the ground. Birds that creep up trees or other vertical surfaces have claws for gripping. A heron's claws are long to provide stability for walking in water or on mud and have comblike serrations to assist in preening.

GENERALISTS AND SPECIALISTS

The nature of the environment in which a bird lives largely determines whether it will be a "specialist" or a "generalist" feeder. Birds that live in habitats where conditions are consistent and food is readily available tend to be highly specialized. Because oceanic environments change relatively little from decade to decade, albatrosses are specifically adapted to soaring over the oceans in search of squid and fish. Even when climatic events alter fish distribution, albatrosses, which are accustomed to flying hundreds of miles in search of prey, simply find their food in other areas of the ocean. In contrast, parrots live in tropical jungles, where climate and rainfall, and therefore the fruiting cycles of fruit trees, can vary from one season to the next. Parrots frequently change their diets in response to the fruit and nuts that are available. Even closely related species can vary in the degree to which their diets are specialized. For terns, with their slender bills and sleek bodies, plunge-diving for fish is the only foraging option. Closely related gulls, with their thicker bills and plumper bodies, use a range of feeding methods and eat a greater variety of foods.

↞ **Some songbirds** are exclusively seed-eaters. Most parrots, by contrast, are generalists in their choice of a wide range of fruits, nuts and seeds. Songbirds, however, adapt to their ever-changing habitat by changing the kinds of seeds they eat. At far left, a Greenfinch (top), a Marsh Tit and a Great Tit (bottom) feed on sunflower seeds in a village garden. At left, an African Gray Parrot nibbles at a twig held in its claw. With their opposable feet, parrots can manipulate objects more readily than any other type of bird.

← **Birds that feed in large groups** can be either generalists or specialists. Most gulls (top) forage in many different locations, including garbage dumps, and employ a variety of feeding strategies to scavenge whatever food is most readily available or most nutritious. Roseate Terns (bottom), which mainly plunge-dive for small fish, are restricted to the world's bays, estuaries and oceans.

Modified for flight

The capacity for sustained flight is the factor that has most influenced the behavior and adaptations of birds. Except for a few flightless species, birds are adapted for the flight that suits their habitat choices and food types. The cigar-shaped, streamlined bodies and long, thin wings of birds that fly constantly—such as albatrosses, petrels and shearwaters—are adapted for soaring. They are built to make use of updrafts over waves and strong gales over the ocean. Some hawks have short, wide wings to maneuver through the underbrush in pursuit of small birds. Accipiters, such as Sharp-shinned Hawks, can duck and turn sharply through thick underbrush without breaking a single feather. Even auks and puffins, which fly only awkwardly through the air, have wings that enable them to fly swiftly and elegantly through the water. Internal structures, too, are adapted to facilitate flight. Birds have hollow bones for lightness, air sacs to increase their ability to take up oxygen, fused vertebrae for structural strength and lightness and a strong keel on the breastbone for the attachment of their strong flight muscles. As well, once the breeding season is over, many birds shed such sexual attractants as long and voluminous tail feathers (pheasants), excess parts of the beak (puffins) and lacy plumes and crests (herons, egrets and others).

LIMITED FLIGHT ADAPTATION

There is no single bird that can walk and run well, soar endlessly and swim rapidly underwater. Each of these ways of moving requires its own special adaptations. Birds with long, strong legs and feet for foraging in shallow water cannot soar as albatrosses and petrels do. Birds that remain on the ground, such as rheas, Ostriches and Emus, have long, heavily muscled legs for running from predators or kicking them. These tall, heavy-bodied, flightless birds are adapted to living in open country, where they can easily catch sight of approaching predators.

↓ **Fulmars soar low** over open oceans, their wings set to glide on the updrafts and air currents. They are heavy-bodied like gulls but range much more widely over the ocean.

← **Eurasian Kestrels** are designed for maneuverability. Flying and gliding over open country, they will often come to a sudden halt, hovering in midair over grasses and tangles of vegetation to seek out any insects, lizards, small snakes, frogs and mice hidden below.

↙ **A Great Spotted Woodpecker** leaves its nest hole in a European tree. These largely sedentary birds are not adapted for long-range flight and spend most of their time perched on tree trunks in woodlands. They fly short distances from tree to tree, and make their way in jerky movements around tree trunks.

FEATHER DETAIL

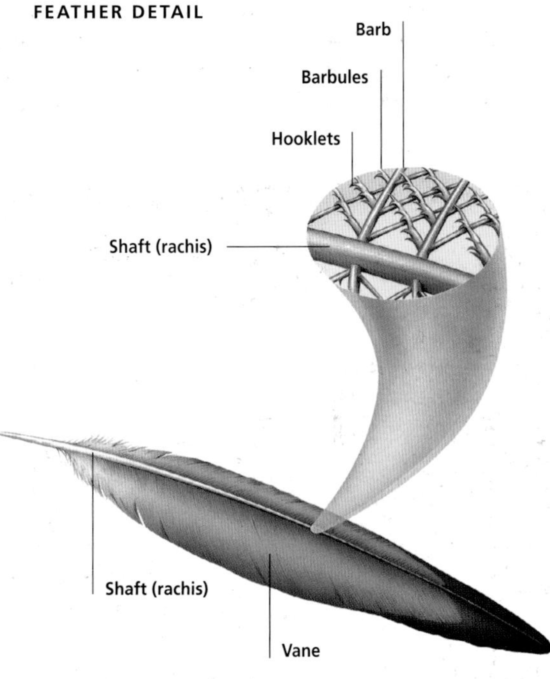

FEATHER STRUCTURE

Although feathers, like the hair of mammals and the scales of reptiles, evolved initially to provide insulation from the cold and heat, wing and tail feathers soon evolved further to permit gliding, and then flying. To be effective for flight, feathers have to be smooth and sleek. The vane of a feather is made up of fine strands that interlock to create a smooth surface. An even row of barbs extends from the feather shaft. Barbules that extend out from the barbs have tiny hooklets which connect with each other to hold the feather together.

Flight patterns

Birds can fly higher, faster and for greater distances than any other animals. They use flight to find food, chase intruders, defend territories, display to mates, find nesting sites and materials, avoid predators or people and migrate to preferred habitats. They also use it to move rapidly from place to place, to soar endlessly on updrafts over mountains or to sail with ocean winds. Birds swoop, soar, glide, dive and undulate through the air. Flight muscles power the wings' stroke, pulling the wings forward and down, and upward and back. The downstroke takes more energy and gives lift; the upstroke is less strenuous. In many species, birds must flap their wings to stay in the air and to move forward. Most birds move their wings continuously up and down and move forward smoothly in a straight line. Other birds save energy by folding their wings close to their bodies for very short periods. This causes them to fall slightly before they resume flapping and again move upward and forward. Birds that fly in this way display an undulating flight pattern; they move up and down rather than straight ahead.

FLYING FROM DANGER

A flock of shorebirds flying from a pursuing predator is one of nature's great spectacles. The mass of birds, packed so densely that they are almost touching each other, wheels and swirls its way through the air. Every few seconds it changes direction, or soars or plunges, in a seemingly synchronized maneuver. In all this movement, birds never fly into each other. There is no leader, and any bird can initiate a change; when one does, other flock members follow instantaneously. In these flights, the shorebirds watch, not the birds closest to them, but more distant members of the flock. This allows individuals to anticipate likely changes in flock direction.

Crows, such as this Carrion Crow, fly by flapping their wings continuously up and down. Feathers streamline the body, making flight easier. Just before crows land, they extend their feet and elevate their wings to brake their speed.

Goldfinches, such as this American Goldfinch, fly in a strongly undulating pattern rather than straight forward. They fly among shrubs and through fields, landing on goldenrods and thistles to forage and gather nest material.

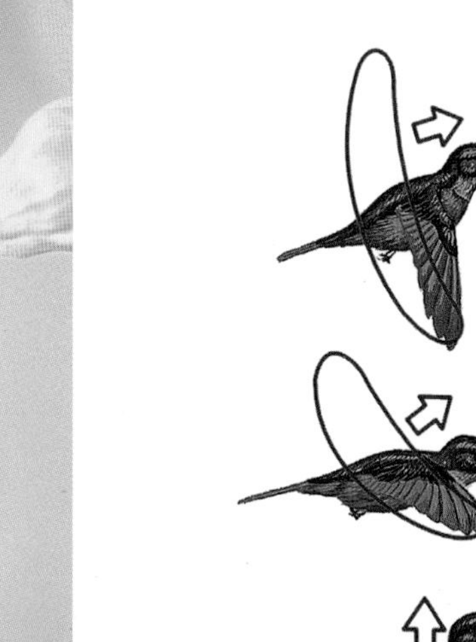

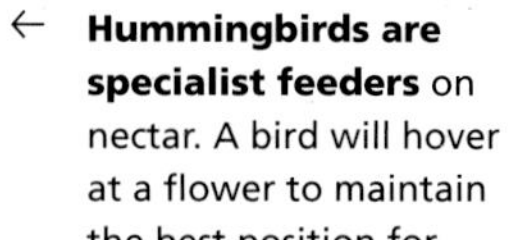

← **Hummingbirds are specialist feeders** on nectar. A bird will hover at a flower to maintain the best position for extracting the nectar. Hummingbirds hover like insects, beating their wings rapidly, then rotating them in different directions at the shoulder joint. This hovering requires a lot of energy. Hummingbirds can also fly forward, backward and straight up or down.

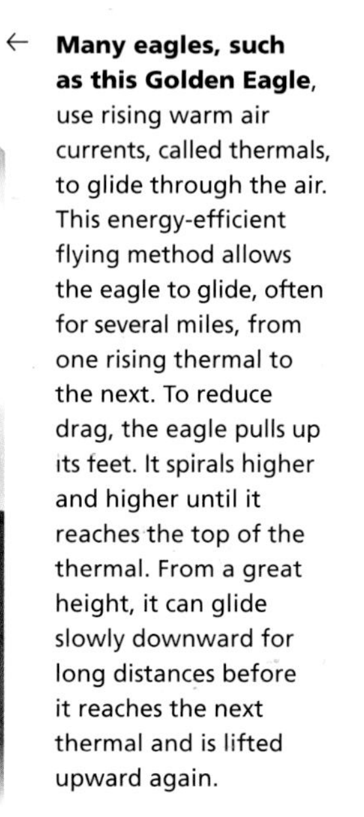

← **Many eagles, such as this Golden Eagle**, use rising warm air currents, called thermals, to glide through the air. This energy-efficient flying method allows the eagle to glide, often for several miles, from one rising thermal to the next. To reduce drag, the eagle pulls up its feet. It spirals higher and higher until it reaches the top of the thermal. From a great height, it can glide slowly downward for long distances before it reaches the next thermal and is lifted upward again.

Life on the ground

Some birds do not fly and live an entirely terrestrial or aquatic life. Flightless birds occur mainly in the Southern Hemisphere, although the recently extinct Great Auk nested in the Northern Hemisphere. Flightlessness evolved separately in many different groups of birds, mainly on isolated islands without mammalian predators. The best-known flightless birds are ratites, a group of birds without keeled sternums. They include the Ostriches of Africa, the Emus of Australia, the cassowaries of New Guinea and Australia, the kiwis of New Zealand and the rheas of South America. The large ratites are mainly strong, diurnal birds of the open country. They rely on long, powerful legs to avoid, or to defend themselves against, predators. Even the small kiwis have powerful legs, but they avoid predators by staying in burrows during the day and foraging at night. Penguins evolved aerial flightlessness rather early, and developed many adaptations for "flying" underwater in pursuit of prey. Since they nested on the low, flat islands in Antarctica, they did not have to fly, unlike Arctic birds, to reach cliff-nesting habitats.

INCREASING DANGERS

Geese, ibises, rails, parrots, cormorants, pigeons, grebes and ducks are some of the bird groups in which flightlessness occurs. Many of these species evolved on small islands without predators, where flight did not improve survival chances. Specialized diving birds, such as the Short-winged Grebe of Lake Titicaca in Peru and the Flightless Cormorant of the Galápagos Islands, have small wings that trap little air and so reduce the birds' buoyancy. These birds survived well until humans arrived and introduced mammalian predators such as cats, dogs and rats. The Kakapo, or Owl Parrot, a flightless bird that lives in New Zealand, hides in burrows and under trees during the day and feeds at night. Several rails, which are limited mainly to small isolated islands, cannot fly. Rats and cats are now eliminating some of these species. The arrival of new predators, including humans, into many habitats has generally made flightless species more liable than flying birds to extinction.

↑ **Most ground-living birds** can fly, although they may not travel far in the air. Northern Bobwhites generally run from danger, but may fly a short distance if the threat persists. They often huddle together against the cold and wind, facing in all directions to look for predators.

→ **Sanderlings forage at the ocean edge**, running down as each wave recedes exposing new foods. They occur along most coasts of the world during migration, although they breed in the tundra of the Arctic. All their foraging time is spent on the ground, either at the leading edge of waves, in back bay mudflats or in tundra pools.

← **Emus stand up to 6 feet (1.8 m) tall** and can weigh 120 pounds (54 kg). Their double feathers give them a shaggy, hairy appearance. When disturbed, they hiss and grunt, and may resort to powerful kicks. Emus and Ostriches have been ranched for their meat, skin and feathers.

↞ **Shy and retiring**, a Great Spotted Kiwi stands quietly in the dense beech forests of New Zealand. These flightless, nocturnal birds specialize in probing the earth for worms and grubs. Nostrils at the tip of the bill allow kiwis to smell prey while they probe.

Night life

Relatively few birds are nocturnal, even though foraging at night means there is less competition with diurnal birds. Owls, nightjars, Oilbirds, Black Skimmers, Kagus, kiwis, thick-knees, Boat-billed Herons and night herons are all nocturnal species that hunt nocturnal prey. Nocturnal skimmers and herons catch fish and shellfish that rise to the surface, and are therefore easier to capture, when it is dark. But sleeping during the day renders birds vulnerable to predators and to the effects of human depredation. Owls, nightjars and potoos overcome this by becoming virtually invisible in daylight. As they sit and sleep quietly on the ground, in trees and on poles, their mottled gray, black and brown coloration blends in with their surroundings. The eyes of birds that are active at night differ from those of diurnal birds. Rods function primarily for black and white vision, and nocturnal birds have more rods than diurnal ones. As well, the vascular membrane, the tapetum, shines through the surface of the retina, giving the eyes a bright red sheen. The acute hearing of some nocturnal species, such as owls, results from the high number of ganglionic cells in the medulla, which process sound and spatial information. Some night hunters have a better sense of smell than other birds: kiwis use olfaction to locate prey; petrels use smell to locate and identify their burrows at night.

ECHOLOCATION

Oilbirds and some cave swiftlets use echolocation—reflected vocalizations—to navigate at night and in their dark caves, and to locate prey. The birds give out pulsed sounds that are echoed back to their ears. Cave swiftlets emit short clicks of a millisecond in duration at normal frequencies (2–10 kilohertz). Oilbirds are fruit-eating nightjars that live in South America. They emit faster clicks, between 15 and 20 milliseconds at 1–15 kilohertz. Oilbirds use echolocation at night but can see in daylight. The echolocation sounds of these birds are in the audible range of humans, and are nowhere near as effective as those of bats.

↓ **Common Potoos** feed at night. Their wide mouths are adapted for catching flying insects. In daylight, their camouflage makes them nearly impossible to see.

← **Thick-knees**, also known as stone curlews, are nocturnal birds that crouch quietly on the ground during the day. Their large eyes help them to feed at night as they socialize noisily.

↙ **Black Skimmers** are most active at dusk and dawn. Some continue foraging throughout the night, especially when both males and females are feeding a hungry brood.

↓ **Eastern Screech Owls** hunt, soon after nightfall, largely for mice. They occur in both red and gray phases. This red-phased bird is clutching a white-footed mouse.

Water birds

Many birds are adapted to aquatic habitats, which include small inland streams and ponds, large lakes and marshes, and coastal bays and estuaries. These environments provide abundant and constantly renewing sources of insects and other invertebrates. Some species—such as grebes, ducks, geese, swans, shorebirds, gulls, dippers and some wrens and blackbirds—live and feed in aquatic environments. They often have thick plumage adapted to insulate and shed water. A number of species build floating nests on the water or attach nests to emergent vegetation. These nests afford some protection from mammalian predators. Floating nests decay from the bottom and must be repaired periodically; nests in emergent vegetation need to be above any floodwaters and firmly attached so that the wind will not dislodge them. Pelicans, herons, egrets, ibises, cranes, gulls and terns usually feed in aquatic habitats but often nest in colonies on dry land or in trees over water. Screamers and some hawks nest in marshes, but feed elsewhere. Screamers feed in the uplands and some hawks prey on birds or small mammals elsewhere. Some birds gain access to a readily available food source by nesting in colonies of other species. Black-crowned Night Herons and Fulmars may nest in mixed-species colonies of gulls and grebes, and eat their neighbors' eggs and chicks.

AQUATIC RISKS

Young that hatch in aquatic nests are at risk of falling into the water, where they could drown or die from cold stress. This is a particular problem for the altricial young of sparrows, blackbirds, wrens and cormorants, which are born with little down and are unable to climb back into their nests. Precocial species, whose young can swim soon after hatching, are less at risk. When they tire of swimming, require comfort or warmth, or need to avoid predators such as snapping turtles, young grebes ride on their parents' backs. The danger of being eaten by neighbors also looms large for chicks that wander alone, far from their nests.

↓ **White Ibises feeding** at sunset probe the mud for crayfish and other invertebrates before going to trees to roost for the night.

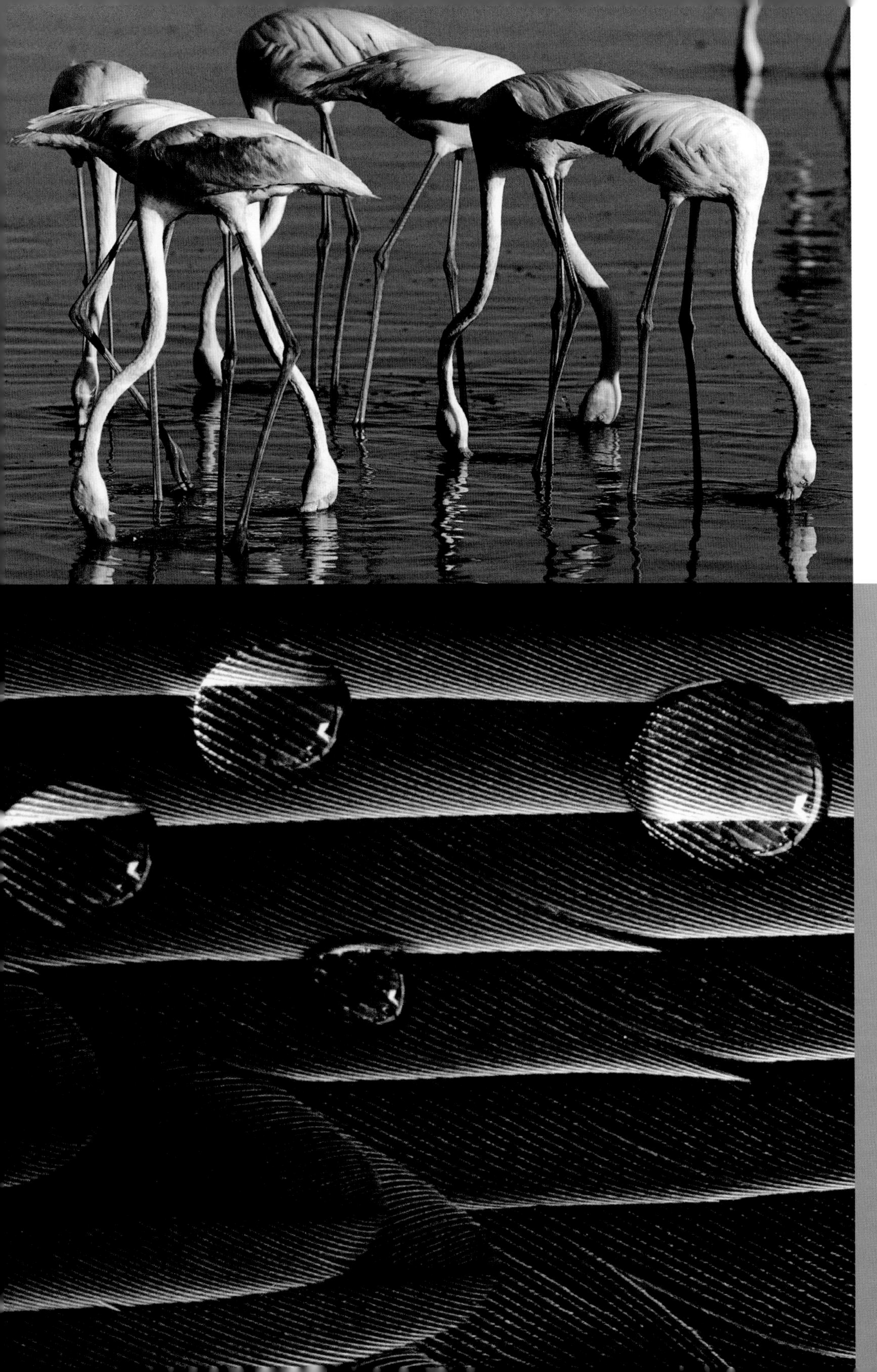

← **Feeding in large flocks** on saline lakes, Greater Flamingos use their oddly shaped bills to strain invertebrates from the water. When feeding, they immerse their heads upside-down underwater and rely on early warning from fellow group members to avoid predators.

↙ **Water beads up** and readily runs off the backs of ducks, such as this Pintail. To protect them from the cold, ducks rely on dense, waterproof feathers and a thick coat of insulating down. They are well adapted to their aquatic lifestyle and spend most of their time on the water.

↓ **When avocets forage** they walk forward or in small circles and swing their upcurved bills back and forth in the shallow water, gleaning small invertebrates. This bird is in its winter plumage.

Shared living spaces

Most birds are social during part of the year—breeding in mixed-species colonies, roosting in large groups and foraging or migrating in flocks. The main advantage of groups is predator avoidance. Individuals in a flock usually have earlier warning of danger than birds that feed or roost alone. A lone bird cannot look in all directions at once, and cannot spend all its waking time watching for predators. Other group advantages include better defense, enhanced foraging, improved selection of migratory stopover areas and the "dilution effect." The dilution effect refers to the reduced chances of being preyed upon in a large group. Most birds roost in inaccessible places, such as in trees, on cliffs or on the water, but some shorebirds, terns and gulls also rest on open beaches, where they are vulnerable to predators. Birds are particularly at risk when they are foraging on the ground. Some birds, such as some crows, Scrub Jays in the United States and Arabian Babblers in Israel, post sentinels at the tops of trees to warn flock members feeding in the shrubbery of approaching predators. During the breeding season many solitary nesters defend breeding territories that are large enough to provide foraging space for themselves and their young. They usually rely on cryptic coloration, secrecy or the calls of their mates or other individuals to warn them of danger.

↑ **Waterfowl commonly roost** in mixed-species flocks because there is enough room to avoid competition for space, and the numbers help deter predators. Even on the water, they are vulnerable to eagles and other large predators. For the smaller ducks, roosting in flocks also facilitates promiscuous pairing each breeding season.

← **Bird feeders** are fast-food restaurants for birds and provide an ideal place to observe birds interacting with members of the same, and other, species. Often dominance relationships are obvious; larger birds usually have first access to food while smaller birds must wait for their turn.

→ **Shorebirds usually breed solitarily**, but migrate in large, mixed-species flocks. Different-sized shorebirds, with varying bill structures, usually feed on slightly different prey items, and this reduces competition.

FORAGING TOGETHER

Mixed-species foraging flocks are common. Sanderlings, Dunlins, plovers and other sandpipers feed in dense flocks on mudflats or beaches; groups of Great Egrets, Snowy Egrets, Little Blue Herons and Glossy Ibises forage in loose flocks in marshes and estuaries; and masses of Red-winged Blackbirds, Brown-headed Cowbirds and Common Grackles feed in dense flocks in marshes or fields, or around cattle. Pastures accommodate mixed flocks of migrating sparrows and buntings. Many species that feed together take slightly different prey items or select seeds or prey of varying sizes. On the African plains and in India, several species of vultures feed together at kills or on carrion. Although the largest species usually has first access to the food, not all vultures are equally adept at breaking open the skin; without larger species present, smaller vultures might not have access to the food at all.

Colonial nesting

Colonial birds squabble about space, kinds and sizes of nest sites, nesting materials and potential mates. Territory size within a colony depends upon the nature of the habitat, the size and type of the birds, the use of the nest site after chicks hatch and the total available space. Birds that nest on cliffs are limited by the number and size of ledges. A ledge must be big enough to accommodate the pair and its young and provide some security. Herons and egrets that build nests in trees have to find branches that can support their weight and are strong enough to protect them against winds. Terns, gulls, skimmers, cormorants, flamingos and other ground-nesters generally colonize a rather flat surface, much of which is suitable for nesting. If chicks leave the nest soon after hatching, territory is usually small. But if the chicks remain until they fledge, then more space is required. Though they nest in big colonies, Herring Gulls usually defend territories because their neighbors will kill any chick that wanders in. Stratified habitats contain higher numbers of nesting birds. Flat-ground habitats can accommodate fewer individuals and species than complex structures such as cliffs, shrubs and trees.

TERN COLONIES

Caspian, Royal, Elegant and Crested, or Swift, terns nest in dense colonies in a hexagonal packing pattern that maximizes the number of nests in an area. Every bird defends only the minimum amount of space needed to care for its nest and chicks. When they all are incubating, they can just touch the bills of their neighbors. Their strategy is to pack as many nests as possible into the available space and to use their territory mainly for courtship and incubation. When the eggs hatch, the chicks remain on the nest for a few days, but then they form crèches that stay together until fledging. Some parents stay with a crèche, but most are off foraging. When parents return with a fish, they call and their chick runs to the edge of the crèche. This assumes early development of individual recognition: chicks recognize the calls of their parents and parents identify their chicks by sight and sound.

↑ **Monk Parakeets in Brazil** sometimes build their nests on the bottom of Jabiru Stork nests. They rely on the storks to keep other predators away.

→ **A Striated Caracara pair** overlooks a rookery of Black-browed Albatrosses and penguins in the Falkland Islands. Caracaras feed on the eggs and young chicks of the albatrosses and penguins.

Migratory adaptations

Even though birds have strong wings, efficient hemoglobin, maximum oxygen uptake because of their air sac system and hollow bones, flying is perilous and strenuous. Birds need a lot of energy to propel themselves through hundreds of miles of air, and fat provides the necessary fuel. As the migration season approaches, birds increase the amount they eat by between 25 and 35 percent, and some eat all day and into the night. This overeating, called "hyperphagia," occurs in all migrants. They often change their diet from insects, which are hard to digest, to fruits, which are easy to digest, high in carbohydrates and can easily be converted to fat. They lay down fat all over their bodies, mainly in layers just beneath the skin, but also in the muscles and organs. Fat is rich in energy; it yields twice as much energy per gram as carbohydrate or protein. The amount of fat stored varies greatly: some birds nearly double their weight during migration, and migrate long distances before stopping; others put on less weight and stop more frequently to eat along the way. Some birds use fats (also called lipids) in a different way. Blackpoll Warblers, for example, are able to regulate their lipid storage in a way that increases the energy it produces. Birds that do this do not need to greatly increase their food intake before migration.

↑ **Migrating Western Sandpipers** gather in dense flocks at the Copper River Delta in Alaska, the largest wetland on the North American Pacific coast.

← **Snow Geese migrate** to the east and west coasts of the United States in winter. They fly at speeds of 50 miles per hour (80 km/hr) and stop periodically to refuel on grasses.

MIGRATION DISTANCES		
Species	**Type**	**Maximum one-way distance**
Blue Grouse	Solitary	1.1. mile (1.8 km)
Black-capped Vireo	Loose straggling groups	1250 miles (2000 km)
Blackpoll Warbler	Loose straggling groups	2480 miles (4000 km)
Painted Bunting	Solitary	3000 miles (4800 km)
Wood Thrush	Loose flocks	3750 miles (6000 km)
Scarlet Tanager	Solitary or small flocks	4350 miles (6960 km)
Bobolink	Dense flocks	4960 miles (8000 km)
Purple Martin	Flocks	6000 miles (9600 km)
Lesser Yellowleg	Flocks	9300 miles (14,880 km)
Red Knot	Dense flocks	10,000 miles (16,000 km)
Arctic Tern	Flocks	11,000 miles (17,700 km)

SAVING ENERGY

Some birds, when they migrate, fly in formation. Others fly in flocks that may be dense and tightly organized, or loose and random. The V-shaped formations of geese and cranes and the single-file patterns of pelicans and cormorants are commonly observed in the sky. These formations help the birds to conserve energy, because the rising, swirling air currents created by the wing tips of the leading birds (called tip vortex) provide lift for the birds that follow. Scientists estimate that flying in V-formation increases geese's energy efficiency by about 20 percent. Birds take turns to lead the formation. Birds that fly in a straight line save energy because the lead bird creates a low-pressure area behind itself. Air moving over the first bird swirls downward, and creates an air current that pulls the second bird forward. Other birds save energy by flying when wind conditions are favorable, rather than against strong gale winds. But even without these energy-saving behaviors, birds can fly thousands of miles without stopping. Birds that fly in flocks usually migrate over greater distances, and have longer continuous flights, than birds that migrate individually.

Seasonal changes

The seasons change everywhere on Earth, although the differences are more extreme in polar than in tropical regions. The drastic seasonal changes in temperature, winds and the amount of sunlight that are typical of polar regions limit the amount of food that is available. Birds cope, either by migration or irruptive movements away from their region, or by finding ways to reduce their food needs while still staying warm. Similar, but less severe, seasonal changes occur in temperate regions, where a higher proportion of bird species remains all winter. Most temperate species, however, still migrate some distance. In tropical regions, too, food varies between the wet and dry seasons. This forces some birds to move or to change to foods that are more readily available, such as different, or lower quality, fruits or nuts. Breeding biology accounts for some of the changes that birds undergo at different times of the year. During the spring and summer breeding season, many birds molt into brightly colored coats, grow long plumes or exotic tail feathers, or acquire other body adornments to attract the opposite sex. After the breeding season they shed these now superfluous appendages and revert to a streamlined and dull-colored form that is more conducive to effective flight and avoiding predators.

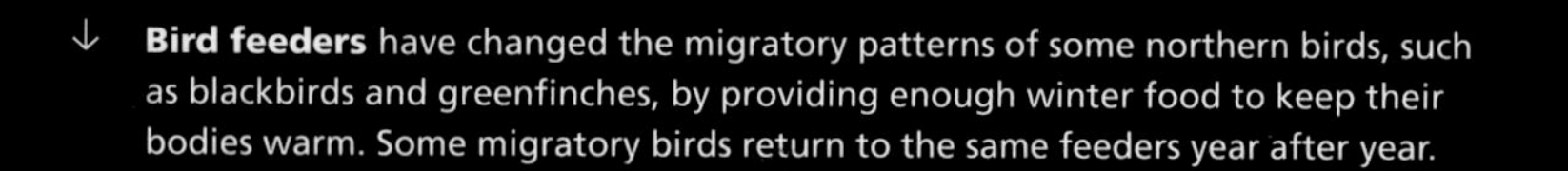

↓ **Bird feeders** have changed the migratory patterns of some northern birds, such as blackbirds and greenfinches, by providing enough winter food to keep their bodies warm. Some migratory birds return to the same feeders year after year.

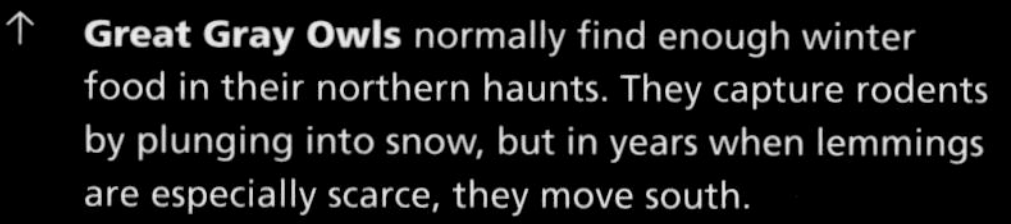

↑ **Great Gray Owls** normally find enough winter food in their northern haunts. They capture rodents by plunging into snow, but in years when lemmings are especially scarce, they move south.

ANNUAL CYCLES

Most birds exhibit cycles that last approximately a year. Birds' internal cycles are thus roughly synchronized with the seasonal cycle of Earth. Changes in day length, or "photoperiod," largely maintain and control this annual cycle. Also significant are internal mechanisms that operate through a complex network of hormonal and brain chemistry interactions. This is often referred to as an internal, or biological, clock. When some birds, such as European Starlings, Garden Warblers and Blackcaps, are kept in laboratories where periods of light and dark alternate at exact 12-hour intervals, they still come into breeding condition at the normal time of year. The pineal gland, located at the base of the brain, is generally regarded as the photosensitive organ in birds, and the location of the biological clock.

→ **Many species have bright breeding plumage**, which they shed during the winter season. Red-headed Weavers exhibit sexual dimorphism, where the males and females have different color patterns. The females are dull colored all year, while the males molt into bright red plumage before the breeding season (right), and revert to dull plumage in the winter (far right). Red-headed Weavers, birds of bush or open woodland country, nest solitarily rather than in groups.

Adjusting to new environments

Birds live predominantly in terrestrial environments, which are subject to seasonal and long-term changes. The slow rise in sea levels, for example, is altering coastal vegetation and the structure of islands and beaches on which many species nest. Birds respond to such changes by finding new habitats, either in the same general region, or farther afield. Changes in climate or ocean currents affect the reproduction patterns and the location of fish populations, altering the times when fish can be preyed upon. Birds that are traveling, either in irruptive food-seeking journeys or because they are expanding their ranges, must adapt to new conditions. Over the last century Northern Hemisphere gulls have extended their range southward in Europe, Asia and North America. Once they left their cliffs and rocky islands in Arctic and other northern habitats, they had to learn to nest on sandy beaches, on sand dunes and in salt marshes. Northern Hemisphere chickadees and titmice, as well as crows and jays, are shifting from woodlands and forests to gardens, parks and suburban woodlots. In these locations, they feed on new kinds of seeds, which they store for later use, and rely on backyard bird feeders. Some birds, such as Peregrine Falcons in the United States, have been introduced to new habitats because their old haunts are no longer available or safe. They now nest on city buildings, bridges and TV towers, as well as on cliff ledges along rivers.

↗ **The loss of native forests** has forced some Ural Owls in the northern Palearctic to use stick nests instead of tree cavities, and to move from boreal forests into more southerly temperate forests.

→ **Starlings, such as this Cape Glossy Starling** at Etosha National Park in Namibia, readily adapt to humans and soon learn that water taps are more reliable, and safer, than waterholes.

SUCCESSFUL MIGRANTS

Birds expand into new locations either on their own, or with human help. Introduced birds may die out quickly or they may adapt by changing habitats, foods and behavior patterns. In South America, Monk Parakeets feed on grass seeds and grain. Individuals that were released in North America quickly adapted to suburban lawns, gardens and backyard feeders, and built large, communal nests on poles and buildings. Eurasian Starlings, which take readily to different habitats and foods, compete with native species for nest holes and breed rapidly, are a classic example of successful adaptation to new environments.

← **Red-billed Queleas** travel nomadically to new environments in huge flocks, searching for the rains that will stimulate their breeding.

↓ **Cattle Egrets**, from Africa, readily adapted to following horses and cows, using them as beaters to scare up insects as they graze.

Adaptive radiation

New species form by adaptive radiation. This is the process by which isolated members of a species change over time until they are sufficiently different from other members of that species that they can no longer interbreed with them. At this point they become a separate species. The evolution, or generation, of new species is called speciation. Adaptive radiation is easiest to observe on island chains, because separate forms of species are isolated on small islands and many generations can pass without any interbreeding. Often a small population of birds colonizes a distant island and becomes isolated, thus cutting off any gene flow (interbreeding) with the parent population. Gradually the isolated population changes and adapts in ways that eventually produce major differences between it and the parental stock. When advantageous mutations occur, they spread rapidly through the population. Natural selection may proceed differently on each island, producing an array of new species. Over time, the founding population on a new island becomes better adapted to its local environment. The best known examples of adaptive radiation on islands are from the Galápagos Islands (finches) and the Hawaiian islands (honeycreepers). Geographical isolation is the key to adaptive radiation, although ecological or habitat isolation can also lead to evolution of new species, especially if there are isolated habitat islands, such as mountaintops.

← **The Woodpecker Finch** of Santa Cruz Island pries insects from bark with its short beak. It sometimes fashions a cactus spine to extract grubs from under bark or cracks in dead wood.

↖ **The Vegetarian Finch** lives on the Galápagos island of Santa Cruz. This tree finch uses its short beak to crush and eat leaves, buds and unripe fruits.

↑ **The Sharp-billed Ground Finch**, from Wenman Island in the Galápagos, is also known as the "vampire finch." It has learned to draw blood from the base of the feathers of Masked Boobies and other seabirds.

↗ **The Cactus Finch**, of Santa Fe Island in the Galápagos, feeds on the pads and fruit of opuntia cactuses. It also collects nectar and pollen.

↓ **Adaptive radiation** is noticeable on the remote Galápagos Islands, where outside influences and competitor species are almost absent.

The Galápagos Islands

DARWIN'S FINCHES

Darwin's finches are known the world over because of the role they played in the formulation of Charles Darwin's theory of evolution. Darwin visited the Galápagos Islands in the 1830s. He observed that one ancestral finch species had apparently diversified and evolved on different islands to take advantage of the varied food supply available. The 14 species of Galápagos finches are a classic example of adaptive radiation. They evolved from a common ancestral finch that reached the Islands from South America. The main divergences between the species are in the types and shapes of the bills, which adapted to suit different foods. There are three basic types of finches: ground finches, warbler-like finches and tree finches. Ground finches have strong, conical beaks that they use to crack and eat seeds; warbler-like finches have thin, pointed beaks with which they pick insects and spiders from flowers, leaves and twigs; and tree finches have strong beaks that can pry insects from leaves, twigs and dead branches. Two species of tree finches, the Woodpecker and Mangrove finches, are famous for their ability to fashion twigs into tools that they use to pull larvae and pupae from holes.

→ **Charles Darwin (1809–1882)** made these sketches of the Galápagos finches during his visit to the Islands in the 1830s. They were reproduced in his book, *A Naturalist's Voyage*.

The human impact

The human impact

For human beings, acquaintance with birds opens a world of esthetic and intellectual enjoyment—but it is a world that is shrinking daily. Birds are threatened by habitat loss, introduced predators and the illegal pet trade. Current conservation measures are helping to reverse this trend.

Shrinking habitats

Habitat loss is the major cause of declining bird populations. Most at risk are habitats that humans want to develop for housing, industry, agriculture or the exploitation of natural resources. Humans clear forests for farmland and wood. They convert grasslands into grain fields, mangrove swamps into shrimp and fish farms and marshes into rice paddies. They put industries along rivers and build towns and cities along coastlines. Birds that are adaptable can survive these disruptions; others cannot. Many cardinals, robins, tits, chickadees, blackbirds, crows, grackles and some parakeets, have adapted successfully to urban environments. Although many changes are the result of natural phenomena, human intervention is dramatically changing the world's landscapes. As well as destroying and degrading many habitats, humans have increased the numbers and kinds of predators in the remaining ones. They have introduced dogs, cats and rats, and caused increases in native predators, such as foxes, by providing food and building roadways into forests.

HABITAT FRAGMENTATION

In many parts of the world the fragmentation of forest habitats has significantly altered the amount, kinds and quality of the habitat available for nesting and foraging birds. When large, continuous areas of forests are divided into smaller and smaller patches, with houses, farms or industrial developments in between, birds no longer have sufficient undisturbed places to breed. Roads through forests not only break up the habitat; they create corridors that provide access to exotic, or non-native, species. When exotic plants move in to a disturbed area, native plant foods often decline or disappear, forcing birds to move elsewhere or to change their diet. Some exotic species prey on native birds; others compete for vital resources. Some species can adapt to the shrinking of forest areas; some are destroyed by them. Habitat fragmentation is also occurring, with a corresponding loss of bird diversity, along sandy beaches, on barrier islands, in chaparral territory and in mangroves.

← **The island of Cebu**, in the Philippines, is an example of coastal depredation. The construction of villages, such as the one seen here, has completely destroyed native mangrove forests and beach habitats used by birds.

↞ **Coastal habitats**, which are home to Arctic Terns and many other species, are shrinking swiftly as a result of many types of human intervention.

← **Destruction of rain forests** is occurring most rapidly in Asia. Here, forest in Borneo is being clear-cut to make way for an oil palm plantation, with a resulting loss in biodiversity.

↞ **Grasslands in many places** are being converted to farmland or grazing lands. Here, a Pied Avocet is defending its nest against an inquisitive, and potentially destructive, cow.

Hostile environments

Humans are the primary cause of declining bird populations. They have disturbed and destroyed bird habitats and introduced a range of contaminants, including toxic chemicals and plastics. While birds are still hunted for food in many places, their exploitation for decorative feathers and for oil has largely ceased. Each year, however, millions of birds die when they collide with buildings, lighthouses, radio and cell phone towers, airport ceilometers and wind turbines. Lights on some of these structures attract birds. Many collide directly with them; others, caught in the light, circle endlessly and die of exhaustion. Songbirds and rails, which migrate at night, are particularly at risk of hitting these buildings. Massive oil and chemical spills can kill countless thousands of birds, which swallow the oil when they preen their feathers. While visible pollution, such as air pollution and rafts of floating plastic, has caused the death of large numbers of birds, persisting low levels of pollution exact a great toll. Although their effects are less obvious, they gradually reduce birds' capacity to cope with their environment, and to reproduce and migrate.

TOXIC CONTAMINANTS

While dead body counts as a result of major catastrophes make for dramatic headlines, toxic chemicals are responsible for a range of sublethal effects that impact on individual survival rates and result in population declines. Low levels of chemicals such as lead, cadmium, mercury and oil affect birds' reproductive function by reducing clutch size and hatching and fledging rates. When parent birds are exposed to toxic chemicals they become less able to display to prospective mates, defend territories, incubate their eggs and care for their chicks when they hatch. Chicks that are exposed to lead and other contaminants are less capable than other hatchlings of hiding from predators, begging effectively or competing with siblings for food. Chicks with low levels of lead have reduced ability to recognize their parents and are more liable to wander into the territories of neighbors, where they may be killed in territorial clashes. Over time, levels of toxic pollutants accumulate and increase, causing ever greater behavioral abnormalities in birds.

← **Oil spills kill** thousands of birds each year. When covered with oil, birds lose their insulation and are unable to swim, fly or forage. As well, oil poisons them when they try to preen it from their feathers.

↓ **Oil digesters** at Nuggets Point on Macquarie Island were used to extract oil from penguins and seals. In the 1880s, King Penguins on this island were almost exterminated.

HUMAN IMPACT

Types of human impact are listed in descending order of the numbers of birds killed each year. The order is approximate as accurate estimates are not possible.

Type of impact	Birds affected
Maintaining pets (cats)	Songbirds, ground-nesting and foraging birds
Communications towers and buildings	Songbirds, night migrants
Oil spills	Seabirds, coastal birds, riverine birds
Long lines and fishing nets	Seabirds: mainly albatrosses and petrels
Hunting	Ducks and waterfowl
Pet trade	Parrots, small songbirds
Aquaculture	Freshwater birds
Farm fields/golf courses	Songbirds, insect- and worm-eaters
Toxic chemicals	Seabirds, coastal birds, freshwater birds
Plastics and floatable items	Seabirds, coastal birds
Egging	Gulls, other seabirds, waterfowl

↞ **Waterfowl of all species**, especially geese and swans, are vulnerable to smoke and air-borne contaminants, which eventually pollute rivers and streams.

← **Plastic debris entangles birds**, limiting their ability to forage, swallow or fly effectively. The head of this Glaucous Gull is caught in a plastic six-pack holder

To the brink and beyond

Birds have existed for more than 150 million years, but it is only in the last few hundred years that massive extinctions have occurred. People are the main culprits. Over many millennia humans have hunted birds and collected eggs for food. As human populations grew, this exploitation became more acute. When people arrived on previously unpopulated islands they often hunted birds that had evolved in the absence of predators. Dodos, Moas, Passenger Pigeons and Eskimo Curlews were all hunted to extinction. The flightless Great Auk was one of several species that were killed to make oil for lamps. Today, introduced predators, destroyed or degraded habitats and environmental pollution—all the result of burgeoning human populations—are taking their toll on many species. Predators such as cats, rats, mongooses and ferrets, that have been brought into areas where they did not exist before, are the greatest destroyers of birds. Trade in pets, especially parrots, also accounts for many extinctions and population declines.

ISLAND PERILS
Birds that evolved on islands without mammalian predators have suffered the greatest extinction rates. When cats, rats and other predators arrived and began preying on eggs, nestlings and fledglings, these birds had no adaptations to cope with them. Sometimes predators were deliberately introduced; more often they escaped into the wild or arrived hidden on ships. Mass extinctions have occurred on the Hawaiian Islands, in New Zealand and on smaller islands, such as Guam. Until the 1960s, native bird populations on Guam were intact. Since then, the rapidly spreading brown tree snake has driven many of them to extinction, and others to the brink.

↙ **A Madagascan man** holds an extinct Elephant Bird's egg. These eggs weighed up to 27 pounds (12 kg).

↓ **Dodos were large**, flightless pigeons from Mauritius that became extinct during the seventeenth century. The parrot in this painting is probably fictional.

→ **Some 70 million years ago** several species of *Hesperornis*, a toothed fish-eater, roamed the seas. They disappeared before dinosaurs became extinct. Passenger Pigeons in North American forests were destroyed by market hunters in the late nineteenth century. Moas, large flightless birds of New Zealand, were hunted to extinction 200 years ago.

Hesperornis

Passenger Pigeon

Moa

To the brink, but with hope

Extinctions of bird species are occurring more quickly and on a larger scale than ever before. Recognizing endangered species, compiling data on their ecology and behavior, and identifying the causes of rapid population declines are essential corrective steps. Also important are the banning of hunting and other forms of exploitation, the control of predators and human intrusions, and mitigating the effects of toxic substances. Management strategies may include captive breeding and transplantation. Because chemicals remain in the environment, and in the food chain, for a long time after they are no longer used, pollution control is a difficult issue. It was several decades after DDT was banned before populations of fish-eating Ospreys, Brown Pelicans, Bald Eagles and Peregrines began to increase. Reducing the emission of mercury from coal-fired power plants would reduce the amount of mercury in the food chain, and ultimately lower the deleterious levels in birds, as well as humans. Habitat destruction can be reversed only if remaining habitats are properly managed, new habitats are created and reserves are established. There are several examples of species whose populations declined to fewer than 50 individuals, but which may now be rescued from near extinction. Among them are the Black Robin of New Zealand, the California Condor, the Whooping Crane of the United States and several species of Caribbean parrots. California Condors were victims of illegal shooting and lead poisoning. Despite public opposition, the last six California Condors were taken into captivity, where they were able to breed successfully before being released into safe locations.

THE PET BIRD TRADE

Public involvement is the key to species conservation. The preservation of endangered species requires the spending of public money and widespread support for the conservation of habitat and the protection of birds from poaching and other forms of human disturbance. Local support is essential, as local people can apply peer pressure to prevent poaching. The pet trade threatens the survival of many tropical parrots, which are the world's most endangered group of wild birds. While the Convention on International Trade in Endangered Species (CITES) has reduced this threat, many poachers circumvent the laws. Between two and five million birds every year are moved from tropical habitats to living rooms in Europe and other affluent countries. More than half are finches, mainly from Africa. Another 43 percent are parrots. Most parrots taken from the wild are exported illegally. Captured and smuggled parrots suffer three times their normal mortality rate. Some rare parrots, such as the Hyacinth Macaw and Spix's Macaw, which is now extinct in the wild, command prices up to US$50,000 each. The issue is complex, because pet birds can foster a more general appreciation of birds, and this often results in a commitment to their conservation. Ideally, pet birds should be bred in captivity, and many are. Birds reared in captivity make better pets than those caught in the wild, because they adapt much more readily to people.

→ **By 1941, hunting and habitat loss** had reduced the population of Whooping Cranes to only 16 individuals. Now, thanks to efforts in Canada and the United States, there are more than 300 of these birds in the wild.

← **At one stage exposure to pesticides** and lead brought numbers of California Condors to fewer than 10. These birds were taken into custody, where they could breed, and 130 now live in the wild. Here, a puppet is used to handle a chick. This is to prevent the chick from imprinting on its human carers.

← **Although once common throughout New Zealand**, the Shore Plover now lives only on South East Island (Rangatira), in the Chatham Group. Numbers of this small wading bird are slowly growing.

←← **The population of the Nene Goose**, the state bird of Hawaii, declined to just 50 in the 1950s. There are now about 500 birds, but they need constant protection from predators.

Introduced and exotic birds

The area, or areas, that a species occupies constitutes that species' "range." Most birds stay within their normal range, although over time a number of factors, including habitat change, competition from other species or climatic variations, may cause this range to expand or contract. Sometimes humans, intentionally or by accident, bring birds into new and unfamiliar territories. Introduced birds, also called "exotics," include game birds such as quail, pheasants, doves and waterfowl. Climate, competition or predation defeat most introduced species. But many thrive, often at the expense of native species. European settlers in New Zealand and the Hawaiian Islands, for example, brought familiar European garden birds to these distant shores, and in the process inadvertently introduced disease and competition. Some species, such as the House Sparrow, were intentionally released because they were thought to be attractive. Doves, bulbuls and sparrows have flourished as exotics in climates that were favorable to them. So, too, have pet birds, such as parrots, mynas and some songbirds, that have been released accidentally. Other birds expanded their ranges without human help and quickly became established. Cattle Egrets, for example, which are native to Africa, have spread rapidly throughout Europe, Asia and the Americas.

SUCCESSFUL EXOTIC SPECIES

Introduced species that thrive are often those, such as parrots, sparrows and starlings, which can adapt to a wide range of habitats and can coexist with humans. In the 1800s, acclimatization societies introduced birds and other animals to many countries. In 1890, after several failed attempts, 60 European Starlings were released in New York City's Central Park, and prospered. By the mid-1900s, starlings had spread across North America and into South America. They nest earlier than native species, taking over their nesting cavities. Exotics are generally more aggressive than native birds and compete more successfully for nest sites. They also tend to thrive in locations where native species are under attack and vulnerable to habitat loss. In some places where mammalian predators have taken a toll on native birds, exotic species have been able to flourish and fill the niches that the natives have left. Visitors to Hawaii, New Zealand or New York City are more likely to encounter exotic than native bird species.

↑ **Starlings darken the sky** at dusk over Brighton, England. They are native to Western Europe but have been introduced into other countries. In the United States they have reached as far as Alaska.

← **Common Mynas** have been introduced in many places and have reached others on their own. They nest in holes in trees, and are common in both urban and suburban areas.

⇇ **During the twentieth century**, Cattle Egrets spread from Africa into Europe, Asia and the Americas. They nest in mixed-species colonies of other egrets and herons, often competing successfully with native species for nest sites. An adult Cattle Egret is at the bottom right of this photo. It has an orange crown.

→ **House sparrows** ultimately thrived after they were introduced, amid considerable controversy, into the United States. They adapted well to agricultural areas, where they ate horse feed and horse droppings, but their numbers have declined in urban regions.

Human predation

Humans have always hunted birds—for food, for oil, for sport, for feathers for decoration and ceremonial use, for down to provide warmth and for museum specimens. As human populations grew and weapons became more sophisticated, birds were killed in increasing numbers. Particularly vulnerable were large ground-dwelling species, such as Moas and Dodos, that lived on islands. These birds were easy to capture and kill, and were large enough to provide abundant food. Market hunting—the killing of birds for commercial use—led to over-exploitation on a grand scale, and particularly threatened migrant shorebirds and waterfowl. In the 1880s, birds were killed to provide feathers for women's hats. One Hawaiian honeycreeper, the Hawaii Mamo, whose yellow rump feathers were fashioned into royal gowns, quickly became extinct. Commercial and recreational hunting in North America drove the once abundant Passenger Pigeon, and probably the Eskimo Curlew, to extinction. While hunting led directly to the disappearance of some species, it seriously depleted populations of many others. Even today, ducks, grouse, quail and shorebirds are widely hunted for recreation, as well as for food. In Asia, swiftlet nests are collected for soup. Songbirds and parrots are captured for pets, and birds of prey are taken for falconry. Scientific collection, however, has greatly declined in recent years.

→ **Northern Europeans** have long exploited Atlantic Puffins for food and eggs. In Iceland, a person on the cliff kills birds and tosses them to a man in a boat below.

↓ **For some parts of the year**, people in the now deserted settlement on St Kilda, Scotland, depended upon the Atlantic Puffin and other seabirds for food.

↑ **From the 1800s to the mid-1900s**, hunting parties, such as this one in Yorkshire, England, slaughtered birds without regard to population numbers and often posed proudly for pictures. Conservation efforts gradually raised awareness of the devastating effect of unlimited hunting.

↗ **In the 1890s**, and in the early years of the twentieth century, models regularly showcased feathered hats that were eagerly sought after by fashion-conscious women. Birds and animals at this time were considered fair game for clothing and other accessories.

→ **Greenlanders used long-handled nets** to catch Little Auks (Dovekies) on their nesting slopes in Narsarsuaq. Faroe Islanders use similar nets to catch flying puffins. This technique is still used by researchers to catch Sooty Terns and other seabirds for banding.

↓ **This pie chart** shows the relative relationship between human-induced problems, in millions of birds killed each year.

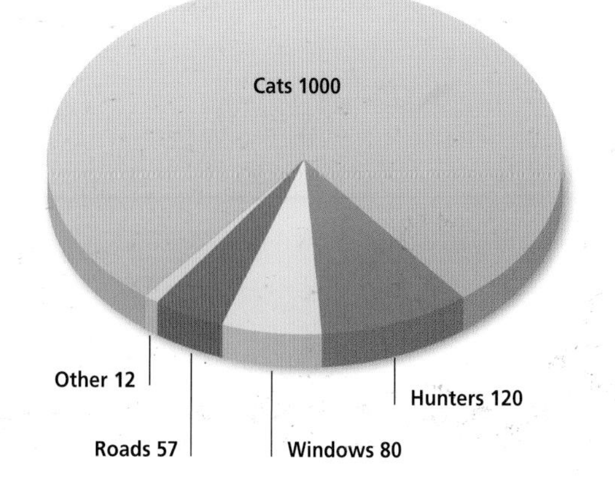

CONTROLLED EXPLOITATION

Human predation is sustainable only when there are enforceable rules in place. Toward the end of the nineteenth century in Europe, the United States and elsewhere, the use of bird feathers, and even whole birds, as adornments for women's hats was considered high fashion. As a result, egrets and herons in particular, but also terns, crows, grouse and songbirds with showy plumes, were massively exploited. Heron and egret populations declined to such an extent that laws were enacted to protect the birds from extinction. Great Auks were targeted for their eggs and their oil. Birds-of-paradise and other birds with spectacular tails have been almost exterminated in places where tribal restrictions on exploitation have broken down.

Recreational hunting is sustainable if governments are able to assess population numbers and regulate the kill. In many places where people collect bird eggs for food, local laws or traditions designed to protect the birds have succeeded in keeping populations stable. In many northern communities, designated families "owned" a section of seabird cliffs and had the sole right to harvest eggs from these birds. Since these families wanted to preserve eggs for future generations, they often took only part of a clutch, early in the season; the birds could still lay more eggs and so raise offspring. Records of egging in some gull colonies, such as those of the Black-headed Gulls at Ravenglass in England, go back hundreds of years.

Bred for consumption

Not content merely to hunt birds, humans soon learned to domesticate a wide range of species for food. These included chickens, geese, ducks, swans, turkeys, pheasants and guineafowl. Pigeons were captured and domesticated first for food, and then as message carriers. However, relatively few of the world's nearly 10,000 species of birds have been domesticated and bred for food or as couriers. The wild Jungle Fowl of Asia is the ancestor of all barnyard chickens. The Chinese domesticated chickens from as early as 3000 BC, and pheasants soon after that. More than 70 breeds are now cultivated for egg production, food, fighting ability or visual appeal. Selective breeding improves the frequency and length of egg-laying, produces bigger eggs, and results in heavier birds or birds whose meat is very tender. The cultivation of ducks, chickens and turkeys for food is still big business. Chickens currently outnumber people by about 20 percent. The tombs of Egyptian kings, dating from 3000 BC, have depictions of overfed captive ducks and geese. These birds were force-fed till they were too plump to be able to walk. The force-feeding of geese continues today, and the excessively fat liver is harvested and sold as foie gras. Soon after they arrived in North America, European settlers domesticated the American Wild Turkey, which has now become a symbol of Thanksgiving.

→ **Modern chickens** are descended from the Red Jungle Fowl, which still lives in Southeast Asian forests. Rhode Island Reds closely resemble the wild ancestral species.

↓ **Today, despite opposition** from animal rights groups, huge numbers of chickens are raised in crowded factory farms.

The familiar Pekin Duck is smooth-headed, but these Crested Pekin Ducks, from a farm in Florida, were originally bred in Asia to provide fluffy wigs for adornment.

In numerous parts of the world, domestic turkeys are still raised on small farms and kept in pens. Many turkeys, however, are cultivated in factory farm conditions.

FACTORY FARMING

During much of the twentieth century, chickens and turkeys on farms were housed in barns or allowed to range freely outdoors. Now, in industrialized countries, chickens are raised on large factory farms where they are confined, often in tens of thousands, in small cages. Today, half of the world's chickens live in these conditions. The United States alone produces more than 74,000 million eggs a year. Chickens that are raised for meat live only a few months before slaughter, so a factory farm might have three "crops" of chickens in a year. At any one time, there are more than 10,000 million chickens in the world; China and the United States each accounts for over 30 percent of this total. Brazil is the other major producer of chickens for meat. Chickens still run free in many less industrialized countries, and are produced on small farms in others.

Birds in captivity

Zoos and game parks display birds to the public but are also active in conservation and education. While zoos once housed birds individually in small cages, many modern zoos have rain-forest and other habitats that mimic the wild, and their displays feature a diversity of birds that have enough space to breed freely. In some zoos, mixed-species colonies of seabirds demonstrate the complexity of social interactions. Rocky shores and underwater exhibits allow people to watch penguins, alcids and cormorants catch fish. Captivity offers experts and non-experts alike the chance to observe avian behavior, learn about physiology and ecological requirements and study population dynamics. Captive-breeding programs are often conducted with species that are endangered or where adult numbers are known to be small. Such programs can produce young that can be reintroduced in the wild to supplement dwindling numbers. Zoos and game parks provide the opportunity to learn more about the breeding biology and behavior patterns of threatened species and so develop strategies for protecting declining populations, such as Japanese Cranes, in the wild. They have also played a key role in the identification of diseases. In 1991, for example, the sudden death of birds in New York's Bronx Zoo alerted biologists to the arrival of West Nile virus in the Western Hemisphere.

AVIAN RESTORATION PROGRAMS

In many avian restoration programs, birds are kept in captivity both for the purposes of study and to provide eggs and birds that can be released in the wild. In some cases, where pesticides and other contaminants have adversely affected certain species, biologists have been able to replace contaminated eggs with clean eggs taken from stable or captive populations. The parents then successfully raise young that eventually will also breed, and so increase the local population. In other instances, young birds are released, either by being placed in the nests of parents who have lost their clutch, or by "hacking" them in nests where they are able to experience and become used to the new habitat. Hacking occurs when birds are placed, within an appropriate habitat, in nest boxes or on platforms and provided with food.

↑ **Cockatiels**, small crested parrots that live in inland Australia, are bred in captivity for the pet trade. They are popular pets because they are small, easy to keep and can be trained to perch on fingers. Cockatiels in the wild are usually sand-colored to blend with the desert background.

← **Songbirds are popular pets** in Latin America and Asia, and many people keep two or three in a small, ornate wooden cage. They often leave the cages hanging on trees while they go to work. On a warm winter's day in parks in Beijing, China, bird cages can be seen hanging up in rows.

⟵⟵ **In the aquarium in Baltimore**, Maryland, people can observe penguins in a semi-natural habitat. They can watch the birds dive and swim to catch their own food. Zoos play an important role in stimulating interest in birds.

Collaboration with birds

Columbus knew he was approaching land in the Americas when he saw terns foraging over schools of fish. Terns rarely forage far from shore because they feed mainly on small fish, and sailors have long been able to read such signs. Even today, fishing expeditions keep watch for seabird flocks, because commercial species often drive small fish to the surface. For more than 1300 years, people in Asia have used cormorants to fish for them. Cormorant fishing is still practiced in Japan and China. Handlers, who spend time with their birds every day and inherit the right to fish, control 10 to 12 cormorants at a time through the skilled use of leashes. They send all of their birds down together and wait for them to emerge with fish. Because of their extensive air sac system, birds respond much more quickly to chemicals than people do. Miners took caged canaries into mines, knowing that if the canaries remained healthy the air would be safe for humans. In some places, geese and other birds are kept as guards, and warn their owners of intruders.

MANAGING USEFUL BIRDS

Humans actively tend birds that are useful to them. They protect the birds and manage the habitat so they can collect guano, eggs or eiderdown. People have long used the down from eiders to make pillows and blankets and to line clothing. Some searched out eider nests in the wild; others cultivated eiders for down. They cared for the eiders in order to increase their flocks and ensure a good supply of eiderdown. For centuries, guano, or dung, from seabird colonies has been a major source of fertilizer because it contains phosphates, nitrogens and other materials essential for plant growth. In some places, people swept surfaces smooth and paved rocky areas to provide more opportunities for seabirds to nest, thereby increasing seabird populations and guano production. Smooth surfaces also made scraping up guano, which was often done with bulldozers, easier. Colonies that were managed for guano included Cape Gannets in South Africa and Guanay Cormorants and boobies in Peru.

← **Seasonal workers from Andean villages** collect guano on Independencia Island in Peru. Peru was one of the most important sources of phosphate fertilizer made from the droppings (or guano) of nesting seabirds. Guano was once Peru's most lucrative export and the key to its economy.

→ **At night, cormorant fishing boats** slide through the water on the Li River in China's Guangxi Province. Each boat has between 10 and 12 cormorants on leashes. An assistant and a helmsman accompany each fisherman.

← **Female eiders pull down feathers** from their own breasts to line their nests. This helps retain heat for the developing eggs. When they leave the nest, the birds pull the brown mottled down over the white eggs to make them invisible to predators. Indigenous northern people collect down for clothing, blankets and pillows. Here, eiderdown is being collected from a nest in Spitsbergen, in the Svalbard group of islands.

A trade in birds

People's appetite for pet birds supports the trade in tropical species. While growing numbers of pet birds are now raised in captivity, most are still taken from the wild. The main "traffic" is from less developed countries to the United States, Europe and Japan, and there is a flourishing trade in hawks to Middle East falconers. Bird trade is regulated internationally by the Convention on International Trade in Endangered Species (CITES), to which more than 150 nations subscribe. CITES has three lists of species: Appendix I lists and prohibits all trade in the most endangered species; Appendix II includes species that would be endangered by unregulated trade; Appendix III contains birds that member countries require help in regulating. Despite this, illegal bird trading still flourishes. Ironically, most birds illegally traded die before they reach their intended destinations. The harvest, both legal and illegal, in wild birds has contributed significantly to the decline and extinction of numerous species, particularly parrots. As a favored species becomes scarce, the pressure to collect the remaining birds intensifies and the price, and therefore the motivation to catch them, increases.

→ **Red and Green Macaws** bite at their cage as they await illegal export. The pet trade endangers many parrots, especially large, showy species. Each individual Red and Green Macaw has its own unique red facial markings.

↓ **Caged birds**, including many songbird species, are displayed for sale in street markets throughout Asia. These ones, in Hong Kong, are stacked row upon row.

↑ **These chicks, in a market** in Vinh Long, Vietnam, are being sold for their potential to provide eggs and meat. Chickens are a staple source of meat in Asia.

VICTIMS OF THE PET TRADE

Large parrots are declining rapidly or disappearing entirely in the wild because of the global pet trade. At one stage, Imperial Parrots in Dominica dwindled to 60 individuals. Educational and conservation programs helped bring the population back to more than 300 birds. The local beer company adopted the Imperial Parrot as its logo, and schoolchildren wrote stories about it and lobbied newspapers to report on it. Spix's Macaw was less fortunate. This magnificent blue macaw, limited to Bahia in Brazil, declined rapidly before its plight became evident. It disappeared from the wild in 2000. There are some Spix's Macaws in captivity, but efforts to coordinate a captive-breeding program have been unsuccessful. To develop survival strategies, parrots rely heavily on social interactions and on skills learned from their parents, and these are difficult to teach in captivity.

Studying birds

Methods of studying birds range from observing their behavior in the wild, and in captivity, to the use of molecular techniques and satellite radio tracking. Observation and experimentation are at the heart of ornithology, and involve the use of telescopes, microscopes, sonographs and laboratory equipment. In their quest for knowledge, ornithologists propose hypotheses, design field or laboratory studies and collect data to test their hypotheses. Laboratory experiments may be relatively short-term. However, long-term studies of marked individuals in the field are necessary to examine such ecological and evolutionary questions as longevity, lifetime reproductive success and parental investment. Individuals can be marked with bird bands, plastic tags on the wings or tail or, temporarily, with dyes. In species where males and females have similar plumage, sexing requires molecular techniques which can also be employed to assign paternity and examine parental investment. Radar permits the study of migration patterns, and radio transmitters read by satellite make it possible to track migrants from Arctic breeding grounds to Antarctic wintering territories.

↓ **Banding or ringing** involves placing numbered metal bands on birds' legs. This Merlin chick is being banded to determine movement patterns and lifespan.

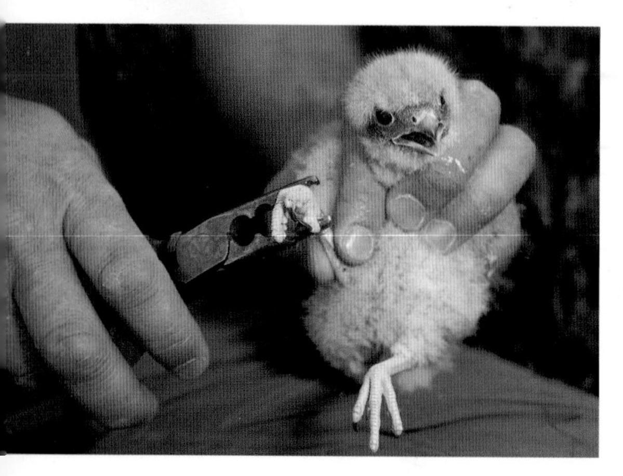

↑ **Seabird researcher Per Anker Nillsen** measures an Atlantic Puffin chick at its roost in Norway. By monitoring birds, researchers can track changes in breeding patterns. They also count the number of wintering seabirds that come to roost each year.

↗ **Charles and Mariana Munn** examine photographs of Red and Green Macaws to see if differences in plumage patterns can be used for individual recognition.

→ **Joanna Burger** is involved in a 30-year study of reproductive success and growth rates, and their relation to contaminant levels in Barnegat Bay, New Jersey, United States. Feathers from this Great Egret chick will undergo metal analysis.

← **By regularly weighing chicks**, such as the Black-browed Albatross, researchers can compare their growth patterns with other chicks in similar or different conditions.

SCIENTIFIC TECHNIQUES

A range of biochemical and molecular techniques is available for studying birds. Scientists use stable isotope analysis to compare the ratios of different isotopes of nitrogen and carbon in birds' bodies with those of food items. The total carbon-to-nitrogen ratio can indicate the quality of a bird's diet. Isotopic research also provides information about the quality of bird eggs formed at different times during the breeding season, and in different geographical areas. DNA (deoxyribonucleic acid) hybridization has been used to determine how closely related different species are, though DNA sequencing is now more commonly used for this purpose. A technique known as polymerase chain reaction (PCR) magnifies minute amounts of DNA from museum specimens. DNA fingerprinting compares the DNA of individuals of the same species to determine parentage, sibling or other relationships.

Conservation measures

Bird conservation is the applied science that employs and integrates information about ecology, behavior, genetics, population biology, exposure to contaminants and the effects of human intervention on birds in order to reverse the widespread decline and extinctions of populations that have been occurring over several centuries. Habitat conservation and restoration, the reduction or eradication of hunting and killing, and the control of introduced or exotic predators are the three most important aspects of avian conservation. The species that are disappearing most quickly, and the reduced habitats on which they depend, demand the most urgent attention. Habitats that are particularly threatened include wetlands, grasslands, old growth forests and wet and dry tropical forests. Indonesia has the highest number of threatened bird species. Brazil, Columbia, Peru, Ecuador, the Philippines, the United States, China, Mexico and India are next in order of concern. While conservation efforts at a local level often focus on one or two endangered species, they sometimes fail to take into account wider environmental threats. Potentially effective initiatives include captive breeding programs; the movement of eggs, young, or even adults, to safer locations; regulations to reduce the use of contaminants and pollutants; and laws that limit human disturbance and harvesting. Conservation is a global issue that requires international control and cooperation; practical measures depend on local public support and commitment for their success.

↓ **Reserves are critical** for the survival of Common Cranes and other cranes that are endangered because of hunting and habitat loss.

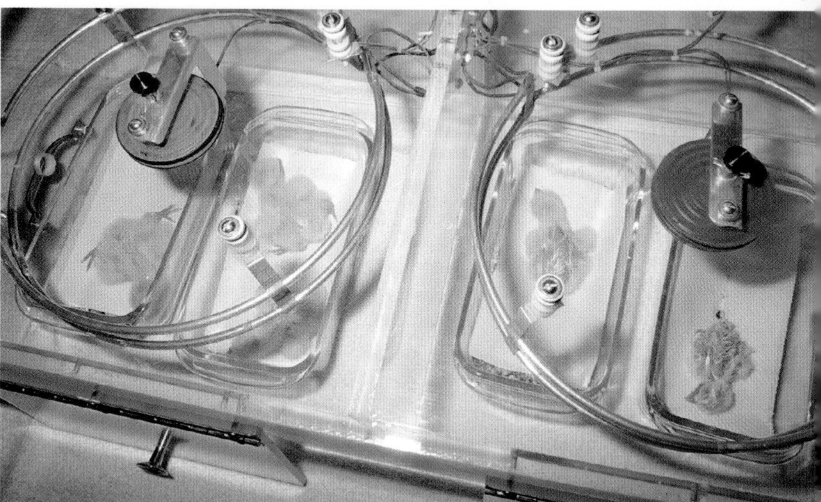

↑ **Pesticides threaten the survival** of Ospreys. Scientists regularly check the condition of Osprey chicks to assess their reproductive success.

↑ **Peregrine Falcons** declined in the 1950s and 1960s as a result of pesticides. Chicks raised in captivity are kept warm in incubators. Captive rearing provides young for reintroduction programs and allows scientists to study the biology of development.

↑ **In some places**, threatened species, such as Peregrine Falcons, rely on large nest boxes for nest sites.

→ **A young, captive crane** is introduced to an artificial parental image to avoid it imprinting on humans.

NATURE RESERVES

People involved in preserving and restoring bird habitats need information, not only about the way birds use habitats, but also about the space they require and the time they need to spend in them. Birds use habitats differently at different times of the year, and reserves must be able to accommodate these differences. The space needed to maintain a viable population will determine the critical size of a reserve. If reserves are smaller than is ideal, corridors between them can sometimes allow birds to move effectively between habitats. Constant monitoring of behavior and breeding patterns within a reserve is essential.

Bird-watching

In the United States alone there are about 61 million bird-watchers, and 20 million people regularly feed birds. Through their purchases of bird-watching equipment and bird-feeding products, bird enthusiasts contribute significantly to local economies. At certain seasons, various communities depend upon bird-watchers to fill their hotels and restaurants and to patronise other businesses. The availability of lightweight, inexpensive binoculars and telescopes enhances both the pleasure and the value of bird-watching. Many birders engage in Christmas bird counts which provide information for tracking winter bird populations, including declines and irruptive movements. In many other ways, as well, amateur bird-watchers gather valuable data that would be impossible to obtain otherwise. Bird enthusiasts range from hard-core "listers," intent on listing as many species as possible, through active encouragers and feeders of garden birds, to backyard birders who watch birds casually. Many tour companies that cater to hard-core bird-watchers organize trips that competitively maximize the number of species that patrons are likely to observe. Birding associations keep track of birds sighted by their members and publish annual lists. A growing number of companies also cater for the needs of casual birders.

→ **A party of tourists** braves cold temperatures and drizzling rain to view a huge colony of King Penguins on South Georgia Island, in the southern Atlantic Ocean.

↓ **Both researchers and amateur bird-watchers** often use field guides, with their excellent illustrations or photographs, to identify birds in the hand.

↟ **Bird blinds**, also called hides, make it possible for bird-watchers, photographers and scientists to observe birds at close quarters and record information about them.

↑ **Bird-watching** is a popular hobby in many countries. Large numbers of bird-watchers often gather with binoculars and telescopes to watch an avian event or add a new species to their "list."

THE INFLUENCE OF BIRD GUIDES

The publication of bird guides has helped stimulate interest in bird-watching. In 1934, Roger Tory Peterson published his landmark *Field Guide to Birds in the United States*, which greatly facilitated the field identification of different species. In the 1960s and 1970s there was a proliferation of bird guides for North America, Europe, Japan and Australia. Guy Tudor's illustrations in books on Venezuela and Colombia set a benchmark for a host of new guides on that region. His attention to subtle variations in plumage patterns made it possible to distinguish, for example, between a diversity of look-alike antwrens and flycatchers. Published field guides have proved invaluable for identifying birds in areas where there are few ornithologists and have helped stimulate local interest and government support for bird protection. The publication of identification guides has also led to an increase in bird tours in numerous places. Many guides feature photographs; others present meticulously observed illustrations that highlight species patterns. Experts still argue over which are more useful for purposes of identification.

Attracting birds

Attracting birds to backyards is basically a matter of making available food and shelter from predators and inclement weather. The simplest way to feed birds is to put up bird feeders and to provision the birds daily. You can use feeders to entice birds close enough to study their behavior. Bird baths, or lily and frog ponds, can serve as sources of water for drinking and bathing. Bird baths, in which birds can stand to wash themselves, are particularly attractive in areas where there is little natural water or when there are drought conditions. Landscaping with trees, shrubs, flowers and groundcovers is the best way to improve backyard habitat conditions for a range of bird species. Vegetation should be planted to offer protection from predators and bad weather, as well as to provide nesting, roosting and perching sites, nesting materials and food. It is important to select an assortment of plants that will produce a variety of buds, seeds, nuts, cones and fruits. Insects will arrive of their own accord. Bird feeders need to be located close to shelter and away from plate glass windows, with which birds can collide if they are suddenly disturbed or frightened. Hole-nesting birds, such as wrens, chickadees, tits, bluebirds, woodpeckers, parrots and parakeets, tree creepers and kookaburras, will take advantage of nest boxes. Bare, dead branches can be useful as perches and nest sites.

USING BIRD FEEDERS

Bird feeders attract an unending succession of birds. They permit the enthusiast to observe aggression between closely related birds; competition between different species; and the predatory attempts of small hawks that specialize on feeding birds. The chance to watch bird behavior at close quarters has awakened a new enthusiasm for birds in many previously indifferent people. Birds that use feeders often "trap-line," moving in a predictable way between feeders in different yards. Banding, or otherwise marking birds at feeders, can provide useful information on foraging behavior, competition, daily movements and numbers of avian visitors. It does, however, require training, experience and, in many places, the possession of an official permit. It is not feasible in all locations.

← **Because there are so few dead trees** left in the wild, many species, such as these American Kestrels, rely largely on nest boxes for their nest sites. They breed successfully in these nest boxes. Here, five juveniles are almost ready to fledge.

← **Woodpeckers, such as this Great Spotted Woodpecker**, as well as many other birds, are especially attracted to bird feeding stations. With their long toes and claws, woodpeckers can cling to vertical surfaces and prop themselves up with their tail feathers.

→ **Bird feeders** are an important source of winter food for many birds. Some migratory birds, such as this male Eastern Bluebird, which normally winters in the southern United States or in Central America, may even remain in the north if sufficient food is provided. Feeders can, however, deter some species from migrating, and if food is suddenly unavailable in bitter winter conditions, the consequences can be serious for these birds.

Training birds

Since ancient times, people have trained falcons and raised carrier pigeons, and early paintings from Asia and other places show parrots talking to their masters. Falconry began in the Middle East and Asia at least 4000 years ago. Falcons appear in writings, frescoes and sculptures and on the walls of Egyptian and Persian tombs. Falconry was the sport of the rich and noble classes, and owning hawks and falcons later became a symbol of power in Europe and Japan. The Mongols kept both falcons and eagles; eagles were sometimes trained to hunt foxes, wolves and even antelopes for food. Falconry is still practiced in many parts of the world. Although pigeons were first domesticated for meat about 4500 BC, they were later trained for carrying messages. The Romans used pigeons to carry news of Caesar's conquests back to Rome, and the first reports of Napoleon's defeat at Waterloo reached England by carrier pigeon. Before phones were reliable, South American cattle breeders informed their neighbors by pigeon when their cows came into heat. Homing pigeons have been trained to return to lofts from hundreds, or thousands, of miles away.

A QUESTION OF INTELLIGENCE

The question of whether trained parrots understand what they say has been resolved. Irene Pepperberg's research has shown that the cognitive abilities of some parrots are equivalent to those of three- or four-year-old children. She first taught Alex, an African Gray Parrot, words for a variety of objects. He could later identify some 50 objects, seven basic colors and five shapes. When asked the question "What is the same?" he correctly nominated objects from an assembled group. He also made specific requests, naming what he wanted. Other birds, too, categorize information to distinguish between different food and habitat types and between predators and non-predators. Birds generally exhibit two features of intelligence: the ability to use experience to solve problems; and the capacity to choose, from among many sets of acquired information, the set that will help solve a current problem.

↑ **Since the late nineteenth century**, pigeon racing has grown in popularity and there are many clubs and organizations devoted to this sport. These highly athletic birds need rigorous training and expert care.

→ **In the rugged, mountainous countryside** of Kazakhstan, in central Asia, men still use trained Golden Eagles to hunt small animals, for both food and sport. Thick gloves protect their wrists against the eagles' strong, sharp claws.

↞ **Producing a succession of doves**, either seemingly out of thin air or from within sleeves, pockets or jackets, is a popular magician's trick. To be convincing, it requires, as well as a skilled human practitioner, a number of well-trained avian performers.

← **Many historic paintings** illustrate the sport of falconry. This illustration, from a thirteenth-century manuscript in the Vatican Library, shows the Holy Roman Emperor Friedrich II with his Master of Falconry.

Birds in lore and legend

Almost 40 kinds of birds, most prominently doves, are mentioned in The Old Testament. In the Book of Genesis, for example, Noah sends out a dove to find land after the flood. Early cave and ancient Egyptian tomb paintings depict birds in detailed behavioral interactions. One Egyptian painting shows kingfishers mobbing crocodiles that are attempting to steal their young. Birds feature strongly in the religious beliefs of many cultures. The Raven is prominent in Inuit creation stories; the Egyptians revered the Sacred Ibis; the Mayans considered the Quetzal to be holy; the Delaware Indians believed that a grebe led survivors of a flood to land; the Crow Indians of Montana thought that diving ducks brought up mud from under primal waters to create the land. A number of mythical birds appear in the legends of different cultures. The Phoenix, a mythical bird that undergoes regular rebirth by fire, appears in legends from places as far apart as Egypt and China. Christians adopted the Phoenix as a symbol of resurrection and immortality and Western depictions often show the bird emerging from flames. To enhance their power, various tribal cultures used feathers in headdresses and coats and for ceremonial purposes. The Hawaiians made cloaks, some of which required half a million feathers, from honeycreeper feathers, and the natives of Borneo made similar capes.

↑ **Not surprisingly, falcons figured prominently** in the hieroglyphics on the Temple of Horus, the falcon-headed god, in Edfu, Egypt. In ancient Egypt, falcons were sometimes represented with human heads.

← **By tradition, Mute Swans** in England belong to the monarch, and they have been designated the "Royal Bird." Swans at the Abbotsbury Swannery in Dorset provide feathers for the helmets of the queen's bodyguard.

→ **An ancient Greek myth** tells of the god Zeus, who took the form of a swan to seduce and abduct Leda, the wife of Tyndareus and the mother of Helen. This painting on a Greek vase shows the abduction.

→ **Birds are often depicted** on stamps and coins. The Bald Eagle, emblem of the United States, has also appeared on several of that country's coins.

BIRDS AS SYMBOLS

Birds often have similar symbolic meanings in different cultures. Large raptors are almost universal symbols of power and strength. An eagle represents the Greek god Zeus; Native Americans use eagle feathers in their leaders' headdresses; and the Bald Eagle is the national emblem of the United States. Hoots and moans of nocturnal birds can be symbolic of impending death and doom. Doves generally signify peace, grebes denote madness, swans evoke pride, and geese stand for confusion. Ravens and crows, and other all-black birds, can be seen as threatening, foretelling evil deeds and death. Some birds have dual personalities. In many South American cultures, the ghostly calls of owls predict misfortune; in ancient Greece, owls were a symbol of knowledge and wisdom and warded off evil spirits. Legends often determine how birds will be treated. The Jains, a religious sect in India, consider it an act of great good to release birds; an industry that provides caged birds for release, often in city parks or during special ceremonies, has grown up around the belief. This custom causes problems when people release birds into unsuitable habitats.

↑ **This pendant, in gold cloisonné**, was found in the tomb of Tutankhamen, which was discovered in the Valley of the Kings, Thebes, in 1922. It depicts Nekhbet, the vulture-goddess and protector of Upper Egypt. She was sometimes depicted with the head of a woman and wearing a vulture headdress.

Birds as inspiration or omens

Birds have figured prominently in the art and literature of most civilizations. Ancient societies passed on poems, stories and word-of-mouth tales of the good and bad that birds foretold. In the two epic poems of the Greek poet Homer, "The Iliad" and "The Odyssey," gods are associated with birds: an owl with Athena, the goddess of wisdom; a falcon with Apollo, the god of the sun; and a dove with Aphrodite, the goddess of beauty. Cocks, cranes, swallows, storks, jays, peacocks, nightingales, eagles and crows all figure in Aesop's fables, oral tales from the Greece of 8000 years ago. In the Sukasaptati, dating from eleventh-century India, a parrot tells a young woman a different tale on each of 70 consecutive evenings to keep her from taking a lover while her husband is away. In Shakespeare's plays, crows and ravens are frequently symbols of blackness and evil; doves and swans suggest whiteness and purity; and nightingales evoke the calm and beauty of dawn and early evening. Birds have provided the inspiration for a host of celebratory poems. Notable among them are Shelley's "To a Skylark," Keats' "Ode to a Nightingale" and Emerson's "The Titmouse." In Coleridge's "The Rime of the Ancient Mariner" an albatross is described as "a pious bird of good omen." It becomes a harbinger of misfortune only after the mariner wantonly kills it.

AWE-INSPIRING DANCES

In many ancient societies, birds were believed to be intermediaries between the material and spiritual worlds and inspired a sense of awe. Although ancient peoples killed birds for food and for their feathers, they often did so within a religious framework of respect and deep appreciation. With their beautiful, and often gaudy, plumage, courtship dances and elaborate display flights, birds have inspired an array of dances in different cultures. In many cases these dances were of great cultural and spiritual significance. Various Native American tribes used feathers in headdresses, and also performed eagle dances that they believed would bring them luck in hunting. Iroquois still perform an eagle dance to celebrate friendship, give thanks and heal sickness. Other Native American peoples performed dances that imitated the movements of male prairie-chickens as they stomped around on their lek territories. In the highlands of Papua New Guinea, people still make feathered headdresses from birds-of-paradise and copy these birds' courtship dances. People on Torres Strait islands perform a head-bobbing dance reminiscent of the dance of the Torresian Imperial Pigeon. The iridescent displays of pheasants in the mountains of China were reflected in the dances of the local people, and the courtship dances of the Japanese Crane were copied by the Hainu, of Hokkaido, Japan.

→ **Thunderbird totem poles** of the West Coast of North America usually feature birds and mammals that were imbued with the power to produce rain, thunder and lightning. The Tlingit tribe, of the Pacific Northwest, believe that the eagle-like Thunderbird carries water in a depression in its back and that as it flies this water spills out in downpours of rain.

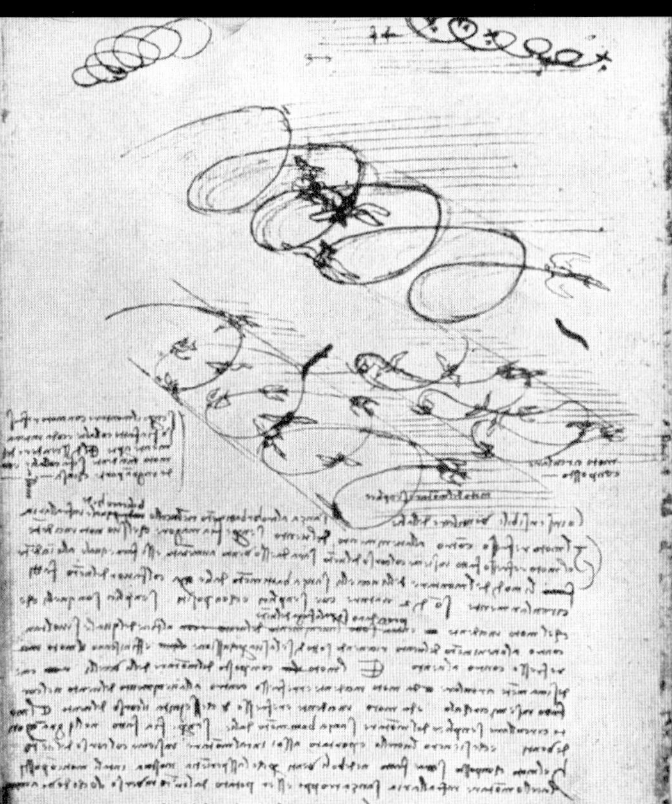

← **Leonardo da Vinci** (1452–1519) was one of the earliest researchers of the flight patterns of birds. The notes that accompany his drawings are in mirror-writing, which Leonardo used to keep his work secret.

→ **This picture, "Mariner, the Albatross,"** shows a frozen ship's crew joined by an albatross. It may have been a source of inspiration for Coleridge's famous poem, "The Rime of the Ancient Mariner."

← **Animals such as birds** commonly featured in petroglyphs—images etched into rock faces—that were created by Native American tribes in the southwestern United States. This one is at Apache Creek, New Mexico.

→ **A white dove in flight**, carrying an olive branch in its bill, is a traditional and widely recognized symbol of peace and goodwill. It has been incorporated into a logo of the United Nations Organization.

Great bird artists

Bird artists have played a significant role in stimulating interest in birds. The invention of the printing press in 1448, and improvements in printing technology, gradually brought bird art to a wide audience. In the seventeenth and eighteenth centuries there was a rash of illustrated natural history works by John Ray, Francis Willughby, Mark Catesby and Thomas Bewick. Their illustrations were often stilted because they were drawn from dead specimens. In the nineteenth century, Alexander Wilson, John James Audubon and John Gould rose to prominence by painting birds that they knew in life. Between 1827 and 1838 an English engraver, Havells of London, produced the four double elephant-folio volumes of *The Birds of America*, which contained 435 Audubon plates with the birds drawn life size. While Audubon concentrated on North American birds, John Gould set out to illustrate the birds of the world and published 41 volumes of illustrations. In the twentieth century many prominent bird artists worked to provide illustrations for field guides, books, magazines or the print market. Some contemporary artists, such as Guy Tudor and Robert Bateman, work in a range of bird art genres.

← **Gould illustrated these Common Black-headed Gulls** in England's Lake District. Here, an unseen predator has caused the chicks to swim out from their nests in the marsh colony. Anxious adults hover overhead or settle on the water near them. Normally these chicks would be hidden in the grass.

→ **An Audubon plate** shows Purple Grackles destroying field corn by opening ears and eating part of them. It was engraved by Robert Havell Jr. and William H. Lizars and published in *The Birds of America*.

↙ **"Great Blue Heron"** was originally a lithograph by the Canadian artist and naturalist Robert Bateman. He painted the bird in its hunting pose as it watches and waits. With lightning speed, the Great Blue Heron can straighten its neck and spear a fish or frog.

TWO GREAT BIRD ARTISTS

The bird drawings of John Audubon (1785–1852) and John Gould (1804–1881) have endured because they are artworks, and not merely illustrations to aid in identification. Although Audubon often painted his birds from collected skins, he observed wild birds and painted them in their habitats, often engaged in typical behaviors. His paintings show puffins in their nesting burrows, prairie-chickens and Horned Larks displaying, Black Skimmers foraging for fish, Barn Owls with prey, Brown Thrashers defending their eggs and Blue Jays stealing other birds' eggs. He painted them with lush vegetation and in the presence of insects, snakes and frogs. Many are beautiful because of their composition and color; some, where he had to squeeze life-sized images into small pages, are stilted and unnatural. The birds in the Havell folios were hand-colored. The Audubon Society of North America was named after him. John Gould also depicted birds in their natural environment. An astute observer of nature, he showed birds nesting, caring for chicks, foraging, stealing and living in colonies. Each painting accurately portrays the birds, their young and the surrounding habitat. Among his most spectacular works are a Great Bustard shielding its chicks from an intruder, large flocks of displaying Common Cranes, a disrupted Black-headed Gull colony, a Common Puffin feeding fish to its downy young, a House Crow stealing eggs from a ground-nesting bird and a displaying male Raggiana Bird-of-paradise. Gould depicted birds of Australia, Asia, Europe, Great Britain and New Guinea.

Factfile

Factfile

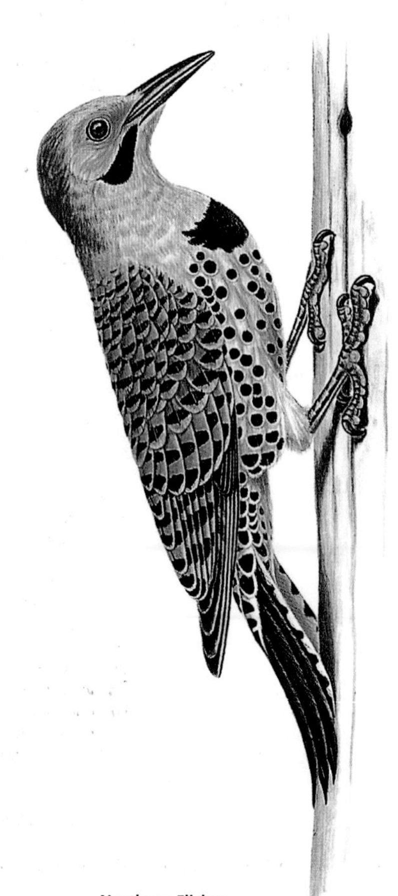

Northern Flicker

BIRD CLASSIFICATION

Order	Family	Species
Tinamiformes	Tinamidae	Tinamous
Struthioniformes	Rheidae	Rheas
	Struthionidae	Ostrich
	Dromaiidae	Emu
	Casuariidae	Cassowaries
	Apterygidae	Kiwis
Galliformes	Cracidae	Curassows, guans, chachalacas
	Megapodiidae	Megapodes, scrubfowl, brush turkeys
	Numididae	Guineafowl
	Phasianidae	Pheasants, turkeys, grouse, quail
Anseriformes	Anhimidae	Screamers
	Anatidae	Ducks, geese, swans
Sphenisciformes	Spheniscidae	Penguins
Gaviiformes	Gaviidae	Divers or loons
Podicipediformes	Podicipedidae	Grebes
Procellariiformes	Diomedeidae	Albatrosses
	Procellariidae	Shearwaters, fulmars, gadfly petrels
	Hydrobatidae	Storm-petrels
	Pelecanoididae	Diving-petrels
Ciconiiformes	Ardeidae	Herons, bitterns
	Balaenicipitidae	Shoe-billed stork
	Scopidae	Hammerhead
	Threskiornithidae	Ibises, spoonbills
	Ciconiidae	Storks
	Phoenicopteridae	Flamingos
Pelecaniformes	Phaethontidae	Tropicbirds
	Fregatidae	Frigatebirds
	Sulidae	Boobies, gannets
	Phalacrocoracidae	Cormorants, shags
	Anhingidae	Anhinga, darters
	Pelecanidae	Pelicans
Falconiformes	Cathartidae	New World vultures, condors
	Pandionidae	Osprey
	Accipitridae	Hawks, eagles
	Sagittariidae	Secretarybird
	Falconidae	Falcons, caracaras
Gruiformes	Rallidae	Crakes, rails, coots, gallinules
	Heliornithidae	Sun-grebes, finfoots
	Rhynochetidae	Kagu
	Eurypygidae	Sunbittern
	Mesitornithidae	Mesites, roatelos

Order	Family	Species
	Turnicidae	Button-quails, Quail Plover
	Gruidae	Cranes
	Aramidae	Limpkin
	Psophiidae	Trumpeters
	Cariamidae	Seriemas
	Otididae	Bustards
Charadriiformes	Jacanidae	Jacanas
	Rostratulidae	Painted snipes
	Scolopacidae	Sandpipers, curlews, woodcocks, phalaropes, etc.
	Dromadidae	Crabplover
	Chionididae	Sheathbills
	Pluvianellidae	Magellanic Plover
	Pedionomidae	Plains-wanderer
	Thinocoridae	Seedsnipes
	Burhinidae	Thick-knees, stone-curlews
	Haematopodidae	Oystercatchers
	Ibidorhynchidae	Ibisbill
	Recurvirostridae	Avocets, stilts
	Glareolidae	Pratincoles, coursers
	Charadriidae	Plovers, lapwings
	Laridae	Gulls and terns
	Stercorariidae	Jaegers and skuas
	Rynchopidae	Skimmers
	Alcidae	Auks, puffins, guillemots, murres, murrelets
Columbiformes	Pteroclidae	Sandgrouse
	Columbidae	Pigeons, doves
Psittaciformes	Psittacidae	Parrots, macaws
	Cacatuidae	Cockatoos
	Loriidae	Lories
Cuculiformes	Cuculidae	Cuckoos, anis
	Opisthocomidae	Hoatzin
	Musophagidae	Turacos
Strigiformes	Tytonidae	Barn Owls
	Strigidae	Typical owls
Caprimulgiformes	Steatornithidae	Oilbird
	Podargidae	Frogmouths
	Aegothelidae	Owlet-frogmouths
	Nyctibiidae	Potoos
	Caprimulgidae	Nightjars, poorwills
Apodiformes	Hemiprocnidae	Crested-swifts
	Apodidae	Swifts
	Trochilidae	Hummingbirds
Coliiformes	Coliidae	Colies, mousebirds
Trogoniformes	Trogonidae	Trogons, quetzals

Order	Family	Species
Coraciiformes	Alcedinidae	Kingfishers, kookaburras
	Todidae	Todies
	Momotidae	Motmots
	Meropidae	Bee-eaters
	Coraciidae	Rollers
	Brachypteraciidae	Ground-rollers
	Leptosomatidae	Cuckoo-roller
	Upupidae	Hoopoe
	Phoeniculidae	Woodhoopoes
	Bucerotidae	Hornbills
Piciformes	Galbulidae	Jacamars
	Bucconidae	Puffbirds
	Indicatoridae	Honeyguides
	Picidae	Woodpeckers, wrynecks
	Lybiidae	African barbets
	Megalaimidae	Asian barbets
	Capitonidae	New World barbets
	Ramphastidae	toucans, aracaris
Passeriformes	Acanthisttidae	New Zealand wrens
	Pittidae	Pittas
	Eurylaimidae	Broadbills
	Philepittidae	Asites
	Tyrannidae	Tyrant flycatchers
	Cotingidae	Cotingas, bellbirds, becards, cocks-of-the-rock
	Oxyruncidae	Sharpbill
	Phytotomidae	Plantcutters
	Pipridae	Manakins
	Furnariidae	Ovenbirds
	Dendrocolaptidae	Woodcreepers
	Thamnophilidae	Antbirds, antshrikes, antwrens, antvireos, bushbirds, fire-eyes, bare-eyes
	Formicariidae	Antthrushes, antpittas
	Conopophagidae	Gnateaters
	Rhinocryptidae	Tapaculos
	Climacteridae	Australian treecreepers
	Menuridae	Lyrebirds
	Atrichornithidae	Scrub-birds
	Ptilonorhynchidae	Bowerbirds
	Maluridae	Fairywrens, grasswrens
	Meliphagidae	Honeyeaters, miners
	Pardalotidae	Pardalotes, bristlebirds
	Acanthizidae	Australian warblers
	Eopsaltriidae	Australian robins
	Irenidae	Leafbirds, fairy bluebirds
	Aegithinidae	Ioras
	Orthonychidae	Log-runners
	Pomatostomidae	Australian pseudo-babblers
	Laniidae	Shrikes, bush shrikes,
	Prionopidae	Helmet-shrikes
	Vireonidae	Vireos, peppershrikes, greenlets, shrike-vireos
	Cinclosomatidae	Quail-thrushes, whip-birds
	Corcoracidae	Australian Chough, Apostlebird
	Pachycephalidae	Whistlers, shrike-tit, thick heads, pitohuis
	Corvidae	Jays, crows, ravens, mapgies, nutcrackers
	Paradisaeidae	Birds-of-paradise
	Artamidae	Wood-swallows,
	Cracticidae	Australian magpies, butcherbirds, currawongs
	Oriolidae	Old World orioles
	Campephagidae	Cuckoo-shrikes, minivets, trillers
	Rhipiduridae	Fantails, Willie Wagtail
	Dicruridae	Drongos
	Monarchidae	Monarch flycatchers
	Malaconotidae	Bushshrikes
	Prionopidae	Helmet-shrikes, batises
	Vangidae	Vanga-shrikes
	Callaeaidae	Wattle crow, Huia, Saddleback
	Picathartidae	Rockfowls, rockjumpers
	Bombycillidae	Waxwings, silky flycatchers
	Dulidae	Palmchat
	Cinclidae	Dippers
	Turdidae	Thrushes, robins
	Timalliidae	Babblers, laughing thrushes
	Muscicapidae	Old World flycatchers, chats, wheatears
	Sturnidae	Starlings, mynahs, oxpeckers
	Mimidae	Mockingbirds, thrashers, catbirds
	Sittidae	Nuthatches, Wallcreepers
	Certhiidae	Creepers, treecreepers
	Troglodytidae	Wrens
	Sylviidae	Old World warblers, wrentits
	Polioptilidae	Gnatcatchers, gnatwrens
	Paridae	Tits, titmice, chickadees
	Remizidae	Penduline-tits
	Aegithalidae	Long-tailed tits, bush tit
	Hirundinidae	Swallows, martins
	Regulidae	Kinglets
	Pycnonotidae	Bulbuls, greenbuls
	Hypocoliidae	Hypocolius
	Cisticolidae	African warblers such as cisticolas, prinias, apalises, etc.
	Zosteropidae	White-eyes AlaudidaeLarks
	Promeropidae	Sugarbirds
	Dicaeidae	Flowerpeckers
	Nectariniidae	Sunbirds
	Melanocharitidae	Berrypeckers, longbills
	Passeridae	Old World sparrows
	Motacillidae	Pipits, wagtails, longclaws
	Prunellidae	Accentors, dunnock
	Ploceidae	Weavers, queleas, fodies, bishops, malimbes, widowbirds
	Estrildidae	Whydahs, waxbills, Zebra Finch, Gould finch, etc.
	Fringillidae	Goldfinches, house finches, rosy finches, redpolls, siskins, crossbills, chaffinches, Pine and Evening grosbeaks
	Drepanididae	Hawaiian honeycreepers
	Emberizidae	New World sparrows, towhees, Old World buntings Galápagos finches, seedeaters
	Parulidae	Wood-warblers and Wren-thrush
	Thraupidae	Tanagers, honeycreepers, flower-piercers, euphonias, etc.
	Cardinalidae	Cardinals, grosbeaks (except Evening and Pine), New World buntings, saltators, Dickcissel
	Icteridae	Orioles, blackbirds, grackles, oropendolas, caciques, cowbirds

Information based on *Bird Families of the World*, cornell.edu

TWENTY CRITICALLY ENDANGERED BIRDS

#	Common name	Scientific name	Main distribution	Reason for critically endangered status
1	Po'o-uli (Black-faced Honeyeater)	*Melamprosops phaeosoma*	Hawaiian Islands (Maui)	Population has declined by approximately 80 percent in the last 10 years. Only two individuals were known to remain but have not been seen since 2004. This species is feared extinct.
2	Cozumel Thrasher	*Toxostoma guttatum*	Mexico (Cozumel Island)	Population decline probably due to boa constrictor introduced in 1971; severely reduced by 1988 and 1995 hurricanes; not seen from 1995 to 2004, when one bird was found. This species has not been seen since Hurricane Wilma in 2005.
3	Philippine Cockatoo	*Cacatua haematuropygia*	Philippines (mainly Palawan)	Extensive loss of lowland habitats; trapping for pet trade.
4	Djibouti Francolin	*Francolinus ochropectus*	Djibouti (found only at two sites)	Hunting and continuing deterioration of habitat has led to a 90 percent population decline in the last 20 years.
5	Galápagos Petrel	*Pterodroma phaeopygia*	Galápagos Islands, ranges to Central and South American coasts	Population has declined by more than 80 percent in the last 60 years, mainly due to introduced predators, such as rats, cats, dogs and pigs. Conservation efforts have stemmed the rate of decline; at least 30,000 pairs survive.
6	Montserrat Oriole	*Icterus oberi*	Montserrat	Long-term habitat destruction, exacerbated by recent volcanic eruptions. Fewer than 800 birds survive in a limited area.
7	Somali Thrush	*Turdus ludoviciae*	Somalia	Almost all of its juniper woodland habitat has been cleared.
8	Lesser Sulphur-crested Cockatoo	*Cacatua sulphurea*	Indonesia, East Timor, Komodo, Sulawesi	Unsustainable trapping for pet trade coupled with habitat loss.
9	Visayan Wrinkled Hornbill	*Aceros waldeni*	Philippines (Negros and Panay islands)	Tiny, fragmented and declining habitat; hunting.
10	White-shouldered Ibis	*Pseudibis davisoni*	Kalimantan and Southeast Asia	Small and rapidly declining, fragmented population threatened by deforestation, draining of wetlands, hunting and disturbances.
11	Asian White-backed Vulture (or White-rumped Vulture)	*Gyps bengalensis*	India to Iran and Southeast Asia	Like other vultures in Asia, this species experienced rapid population decline, eventually attributed to feeding on carcasses of animals treated with toxic veterinary drugs, such as diclofenac; perhaps in combination with other causes.
12	Indian Vulture	*Gyps indicus*	India and Pakistan	As above, toxic drugs in combination with other causes.
13	Slender-billed Vulture	*Gyps tenuirostris*	Kashmir to Southeast Asia (recent split from *G. indicus*)	As above, toxic drugs in combination with other causes.
14	Chatham Island Taiko (Petrel)	*Pterodroma magentae*	New Zealand (only Chatham Islands and surrounding waters)	Long-term historical decline. Fewer than 150 birds breed only on one small island which is 1.2 sq m (3 sq km) in area.
15	Sociable Lapwing	*Vanellus gregarious*	Breeds in Kazakstan and southcentral Russia, winters in northeast Africa to India	Rapid population decline is poorly understood, but is expected to continue in the future. Only about 200 pairs known to survive.
16	Blue-billed Curassow	*Crax alberti*	Columbia (Santa Marta to Tolima)	Rapid deforestation, increased access and hunting; fewer than 2500 birds survive in an area of 810 sq m (2100 sq km).
17	Gurney's Pitta	*Pitta gurneyi*	Myanmar, Thailand	Destruction of remaining forest habitat and trapping for pet trade. On the verge of extinction. Tiny, declining population of 50–100 birds occupies an extremely small and shrinking range.
18	Sulu Hornbill	*Anthracoceros montani*	Philippines (survives only on Tawitawi Island)	Tiny population of 40 birds survives on one island; accelerating loss of forest habitat; imminent extinction.
19	Rudd's Lark	*Heteromirafra ruddi*	South Africa	Habitat loss due to commercial planting of trees on former grassland; however, recent efforts to preserve this habitat warranted a downgrading to "vulnerable" status in 2005.
20	Dark-eared Brown-dove	*Phapitreron cinereiceps*	Philippines (found at four sites on Tawitawi)	This recently discovered species has a tiny range, threatened by a decline in the extent and quality of its forest fragments.

Information from the IUCN *Red List of Threatened Species* 2004 and Bird Life International

BIRD-WATCHING ETHICS

It is important that all bird-watchers are aware of their responsibilities in relation to birds and to the natural environment in general. Guidelines of good bird-watching behavior, such as those embodied in the *Bird-watcher's Code of Conduct,* should be followed by bird-watchers all over the world. The members of the American Birding Association pledge to adhere to the following "general guidelines of good birding behavior."

Bird-watchers must always act in ways that do not endanger the welfare of birds or other wildlife

- Observe and photograph birds without knowingly disturbing them in any significant way.
- Avoid chasing or repeatedly flushing birds.
- Only sparingly use recordings and similar methods of attracting birds and not use these methods in heavily birded areas.
- Keep an appropriate distance from nests and nesting colonies so as not to disturb or endanger them.
- Refrain from handling birds or eggs unless engaged in recognized research activities.

Bird-watchers must always act in ways that do not harm the natural environment

- Stay on existing roads, trails, and pathways whenever possible to avoid trampling or otherwise disturbing fragile habitat.
- Leave all habitat as we found it.

Bird-watchers must always respect the rights of others

- Respect the privacy and property of others by observing "No Trespassing" signs and by asking permission to enter private or posted lands.
- Observe all laws and the rules and regulations that govern public use of birding areas.
- Practice common courtesy. For example, limit requests for information and make them at reasonable hours of the day.
- Always behave in a manner that will enhance the image of the birding community in the eyes of the public.

Bird-watchers in groups should assume special responsibilities

- Take care to alleviate the problems and disturbances that are multiplied when more people are present.
- Act in consideration of the group's interest, as well as the individual's.
- Support by individual actions the responsibility of the group leader(s) for the conduct of the group.
- Group leaders should assume responsibility for the conduct of the group.
- Learn and inform the group of any special rules, regulations, or conduct applicable to the area or habitat being visited.
- Limit groups to a size that does not threaten the environment or the peace and tranquillity of others. Teach others birding ethics by words and example.

PLANTING TO ATTRACT BIRDS

Shrubs and vines		
Elderberry (*Sambucus*)	Fruit, insects, shelter	Warblers, vireos, robins, thrushes, Starlings
Honeysuckle (*Lonicera*)	Flowers, insects	Warblers, wrens, hummingbirds, thrushes
Blackberry/raspberry (*Rubus*)	Fruit, dense cover, insects	Wrens, thrushes, sparrows, thrashers, buntings
Gooseberry/currant (*Ribes*)	Fruit, nectar	Kinglets, thrushes, hummingbirds
Hawthorn (*Crataegus*)	Fruit, cover, insects	Warblers, thrushes
Ivy (*Hedera*)	Fruit, dense cover, insects	Woodpigeon, thrushes, Blackcaps
Trees		
Holly (*Ilex*)	Fruit, dense cover	Thrushes
Oak (*Quercus*)*	Acorns, insects, cover	Wood pigeons, jays, tits, Nuthatches, chickadees, woodpeckers
Pine (*Pinus*)*	Seeds, insects, cover	Crossbills, Goldcrests, creepers, woodpeckers
Maple (*Acer*)*	Seeds, insects	Finches, warblers, vireos, tanagers
Plum/cherry (*Prunus*)*	Fruit, insects, nectar	Warblers, thrushes, tits, Bullfinches, Hawfinches, orioles, sapsuckers
Mulberry (*Morus*)*	Fruit, insects, cover	Warblers, woodpeckers, thrushes, grosbeaks
Yew (*Taxus*)	Fruit, dense cover	Thrushes, crests

* Ensure the right species for the geographical area is used.

FEATHERS AND PLUMAGE

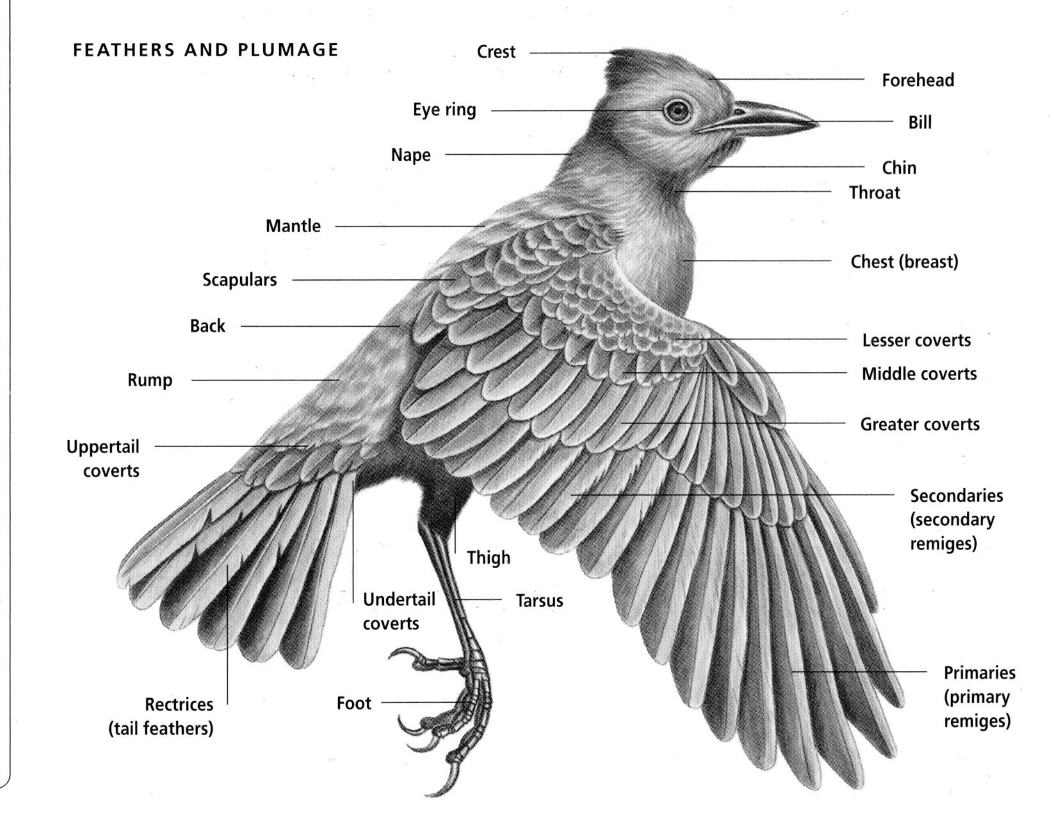

BIRD SIZES

Following is a listing of the height or length of a selection of common birds. Most birds are measured from the tip of their bill to the tip of their tail. Penguins, and some ratites and tinamous, are measured according to their total head and body length. Measurements represent the largest size the average bird in each species can reach.

↕ = Head and body height
↔ = Tip of bill to tip of tail

Common name	Scientific name	Length/height
Ratites and tinamous		↕ ↔
Elegant Crested Tinamou	*Eudromia elegans*	↔ 16 in (41 cm)
Emu	*Dromaius novaehollandiae*	↕ 6.5 ft (2 m)
Great Tinamou	*Tinamus major*	↔ 18 in (46 cm)
Greater Rhea	*Rhea americana*	↕ 5.25 ft (1.6 m)
Little Spotted Kiwi	*Apteryx owenii*	↔ 17.5 in (45 cm)
Ostrich	*Struthio camelus*	↕ 9.5 ft (2.9 m)
Southern Cassowary	*Casuarius casuarius*	↕ 6.5 ft (2 m)
Variegated Tinamou	*Crypturellus variegatus*	↔ 13 in (33 cm)
Gamebirds		↔
California Quail	*Callipepla californica*	11 in (28 cm)
Gray-striped Francolin	*Francolinus griseostriatus*	13 in (33 cm)
Great Argus	*Argusianus argus*	6.5 ft (2 m)
Great Curassow	*Crax rubra*	36 in (92 cm)
Indian Peafowl	*Pavo cristatus*	7 ft (2.1 m)
Koklass Pheasant	*Pucrasia macrolopha*	25 in (64 cm)
Mikado Pheasant	*Syrmaticus mikado*	33.5 in (86 cm)
Ocellated Turkey	*Meleagris ocellata*	4 ft (1.2 m)
Red Junglefowl	*Gallus gallus*	29.25 in (75 cm)
Red Spurfowl	*Galloperdix spadicea*	14 in (36 cm)
Red-legged Partridge	*Alectoris rufa*	14.75 in (38 cm)
Reeves' Pheasant	*Syrmaticus reevesii*	7 ft (2.1 m)
Rock Ptarmigan	*Lagopus mutus*	14.75 in (38 cm)
Vulturine Guineafowl	*Acryllium vulturinum*	23.5 in (60 cm)
White-crested Guan	*Penelope pileata*	32.33 in (83 cm)
Waterfowl		↔
Bean Goose	*Anser fabalis*	34.33 in (88 cm)
Black-necked Swan	*Cygnus melanocoryphus*	3.75 ft (1.15 m)
Canada Goose	*Branta canadensis*	3.75 ft (1.15 m)
Comb Duck	*Sarkidiornis melanotos*	29.66 in (76 cm)
Common Eider	*Somateria mollissima*	27 in (69 cm)
Common Pochard	*Aythya ferina*	17.5 in (45 cm)
Common Shelduck	*Tadorna tadorna*	25.33 in (65 cm)
Coscoroba Swan	*Coscoroba coscoroba*	3.75 ft (1.15 m)
Freckled Duck	*Stictonetta naevosa*	23 in (59 cm)
Magpie Goose	*Anseranas semipalmata*	35 in (90 cm)
Mallard	*Anas platyrhynchos*	25.33 in (65 cm)
Muscovy Duck	*Cairina moschata*	32.75 in (84 cm)
Mute Swan	*Cygnus olor*	5 ft (1.5 m)
Northern Shoveler	*Anas clypeata*	18.75 in (48 cm)
Oldsquaw (Long-tailed Duck)	*Clangula hyemalis*	16.5 in (42 cm)
Orinoco Goose	*Neochen jubata*	25.75 in (66 cm)
Red-breasted Goose	*Branta ruficollis*	21.5 in (55 cm)
Red-crested Pochard	*Netta rufina*	22.66 in (58 cm)
Snow Goose	*Anser caerulescens*	31.25 in (80 cm)
Southern Screamer	*Chauna torquata*	37 in (95 cm)
Torrent Duck	*Merganetta armata*	18 in (46 cm)
White-faced Whistling-duck	*Dendrocygna viduata*	19.5 in (50 cm)
Whooper Swan	*Cygnus cygnus*	5 ft (1.5 m)
Wood Duck	*Aix sponsa*	20 in (51 cm)
Penguins		↕
Adelie Penguin	*Pygoscelis adeliae*	23.75 in (61 cm)
Emperor Penguin	*Aptenodytes forsteri*	4 ft (1.2 m)
Jackass Penguin	*Spheniscus demersus*	3.25 ft (1 m)
King Penguin	*Aptenodytes patagonicus*	3.25 ft (1 m)

Common name	Scientific name	Length/height
Little Penguin	*Eudyptula minor*	17.5 in (45 cm)
Royal Penguin	*Eudyptes schlegeli*	27.33 in (70 cm)
Snares Penguin	*Eudyptes robustus*	23.5 in (60 cm)
Yellow-eyed Penguin	*Megadyptes antipodes*	23.5 in (60 cm)
Divers and Grebes		
Arctic Loon	*Gavia arctica*	26.75 in (68.5 cm)
Common Loon	*Gavia immer*	35 in (90 cm)
Eared Grebe	*Podiceps nigricollis*	13 in (33 cm)
Great Crested Grebe	*Podiceps cristatus*	25 in (64 cm)
Great Grebe	*Podiceps major*	30.5 in (78 cm)
Hooded Grebe	*Podiceps gallardoi*	13.25 in (34 cm)
Little Grebe	*Tachybaptus ruficollis*	11 in (28 cm)
New Zealand Grebe	*Poliocephalus rufopectus*	11.66 in (30 cm)
Red-throated Loon	*Gavia stellata*	27.33 in (70 cm)
Western Grebe	*Aechmophorus occidentalis*	29.66 in (76 cm)
Yellow-billed Loon	*Gavia adamsii*	35 in (90 cm)
Albatrosses and petrels		
Band-rumped Storm-petrel	*Oceanodroma castro*	9 in (23 cm)
Cape Petrel	*Daption capense*	15.25 in (39 cm)
Common Diving-petrel	*Pelecanoides urinatrix*	9.75 in (25 cm)
Gray Petrel	*Procellaria cinerea*	19.5 in (50 cm)
Jouanin's Petrel	*Bulweria fallax*	12.5 in (32 cm)
Northern Fulmar	*Fulmarus glacialis*	19.5 in (50 cm)
Royal Albatross	*Diomedea epomophora*	4 ft (1.2 m)
Wedge-tailed Shearwater	*Puffinus pacificus*	18 in (46 cm)
Wilson's Storm-petrel	*Oceanites oceanicus*	7.5 in (19 cm)
Yellow-nosed Albatross	*Thalassarche chlororhynchos*	29.66 in (76 cm)
Flamingos		
Andean Flamingo	*Phoenicoparrus andinus*	3.66 ft (1.1 m)
Greater Flamingo	*Phoenicopterus ruber*	4.75 ft (1.45 m)
Herons and allies		
Boat-billed Heron	*Cochlearius cochlearius*	20 in (51 cm)
Cattle Egret	*Bubulcus ibis*	20 in (51 cm)
Great Blue Heron	*Ardea herodias*	4.5 ft (1.4 m)
Hamerkop	*Scopus umbretta*	21.75 in (56 cm)
Sacred Ibis	*Threskiornis aethiopicus*	35 in (90 cm)
Whistling Heron	*Syrigma sibilatrix*	23.75 in (61 cm)
White-crested Bittern	*Tigriornis leucolophus*	31.25 in (80 cm)
Wood Stork	*Mycteria americana*	3.33 ft (1 m)
Pelicans and allies		
Anhinga	*Anhinga anhinga*	35 in (90 cm)
Dalmatian Pelican	*Pelecanus crispus*	5.5 ft (1.7 m)
Darter	*Anhinga melanogaster*	37.75 in (97 cm)
Double-crested Cormorant	*Phalacrocorax auritus*	35.5 in (91 cm)
European Shag	*Phalacrocorax aristotelis*	30.75 in (79 cm)
Great Cormorant	*Phalacrocorax carbo*	3.25 ft (1 m)
Great White Pelican	*Pelecanus onocrotalus*	5.75 ft (1.75 m)
Lesser Frigatebird	*Fregata ariel*	31.66 in (81 cm)
Northern Gannet	*Morus bassanus*	36 in (92 cm)
Pelagic Cormorant	*Phalacrocorax pelagicus*	28.75 in (74 cm)
Peruvian Booby	*Sula variegata*	3.25 ft (1 m)
Red-tailed Tropicbird	*Phaethon rubricauda*	19.5 in (50 cm)

Common name	Scientific name	Length/height
Birds of prey		
African Cuckoo-hawk	*Aviceda cuculoides*	15.66 in (40 cm)
African White-backed Vulture	*Gyps africanus*	36.66 in (94 cm)
Andean Condor	*Vultur gryphus*	4.25 ft (1.3 m)
Bald Eagle	*Haliaeetus leucocephalus*	3.66 ft (1.1 m)
Black Baza	*Aviceda leuphotes*	13.75 in (35 cm)
Black Harrier	*Circus maurus*	19.5 in (50 cm)
Black Kite	*Milvus migrans*	21.5 in (55 cm)
Cinereous Vulture	*Aegypius monachus*	3.25 ft (1 m)
Collared Falconet	*Microhierax caerulescens*	7 in (18 cm)
Common Kestrel	*Falco tinnunculus*	14.5 in (37 cm)
Crested Serpent-eagle	*Spilornis cheela*	29.66 in (76 cm)
Crowned Eagle	*Harpyhaliaetus coronatus*	33 in (85 cm)
Dark Chanting-goshawk	*Melierax metabates*	21 in (54 cm)
Egyptian Vulture	*Neophron percnopterus*	27.33 in (70 cm)
Eurasian Griffon	*Gyps fulvus*	3.66 ft (1.1 m)
Eurasian Sparrowhawk	*Accipiter nisus*	14.75 in (38 cm)
European Honey-buzzard	*Pernis apivorus*	22.66 in (58 cm)
Harris's Hawk	*Parabuteo unicinctus*	22.66 in (58 cm)
Hooded Vulture	*Necrosyrtes monachus*	27 in (69 cm)
Javan Hawk-eagle	*Spizaetus bartelsi*	23.75 in (61 cm)
King Vulture	*Sarcoramphus papa*	31.66 in (81 cm)
Martial Eagle	*Polemaetus bellicosus*	32.33 in (83 cm)
Mississippi Kite	*Ictinia mississippiensis*	13.75 in (35.5 cm)
Osprey	*Pandion haliaetus*	22.66 in (58 cm)
Palm-nut Vulture	*Gypohierax angolensis*	19.5 in (50 cm)
Red-footed Falcon	*Falco vespertinus*	12 in (31 cm)
Scissor-tailed Kite	*Chelictinia riocourii*	13.75 in (35 cm)
Secretary Bird	*Sagittarius serpentarius*	5 ft (1.5 m)
Short-toed Eagle	*Circaetus gallicus*	27 in (69 cm)
Snail Kite	*Rostrhamus sociabilis*	16.75 in (43 cm)
Turkey Vulture	*Cathartes aura*	31.66 in (81 cm)
Variable Goshawk	*Accipiter novaehollandiae*	21.5 in (55 cm)
Yellow-headed Caracara	*Milvago chimachima*	16.75 in (43 cm)
Cranes and allies		
African Finfoot	*Podica senegalensis*	23 in (59 cm)
Barred Buttonquail	*Turnix suscitator*	6.66 in (17 cm)
Black Crowned Crane	*Balearica pavonina*	3.5 ft (1.05 m)
Corncrake	*Crex crex*	11.66 in (30 cm)
Demoiselle Crane	*Anthropoides virgo*	35 in (90 cm)
Denham's Bustard	*Neotis denhami*	3.25 ft (1 m)
Hoatzin	*Opisthocomus hoazin*	27.33 in (70 cm)
Horned Coot	*Fulica cornuta*	20.66 in (53 cm)
Lesser Florican	*Sypheotides indica*	20 in (51 cm)
Limpkin	*Aramus guarauna*	27.33 in (70 cm)
Sunbittern	*Eurypyga helias*	18.75 in (48 cm)
White-breasted Mesite	*Mesitornis variegatus*	4 ft (1.2 m)
Waders, gulls and allies		
Atlantic Puffin	*Fratercula arctica*	14 in (36 cm)
Beach Stone Curlew	*Esacus magnirostris*	21.75 in (56 cm)
Black Skimmer	*Rynchops niger*	18 in (46 cm)
Black-faced Sheathbill	*Chionis minor*	16 in (41 cm)
Black-tailed Godwit	*Limosa limosa*	16.5 in (42 cm)
Black-winged Stilt	*Himantopus himantopus*	15.66 in (40 cm)
Collared Pratincole	*Glareola pratincola*	9.75 in (25 cm)

BIRD SIZES CONTINUED

Common name	Scientific name	Length/height
Common Redshank	*Tringa totanus*	11 in (28 cm)
Common Snipe	*Gallinago gallinago*	10.5 in (27 cm)
Common Tern	*Sterna hirundo*	14.75 in (38 cm)
Crested Auklet	*Aethia cristatella*	10.5 in (27 cm)
Curlew Sandpiper	*Calidris ferruginea*	8.5 in (22 cm)
Eurasian Curlew	*Numenius arquata*	23.5 in (60 cm)
Fairy Tern	*Sterna nereis*	10.5 in (27 cm)
Great Black-backed Gull	*Larus marinus*	29.66 in (76 cm)
Herring Gull	*Larus argentatus*	25.75 in (66 cm)
Ibisbill	*Ibidorhyncha struthersii*	16 in (41 cm)
Little Tern	*Sterna albifrons*	11 in (28 cm)
Long-tailed Jaeger	*Stercorarius longicaudus*	20.66 in (53 cm)
Pied Avocet	*Recurvirostra avosetta*	16.75 in (43 cm)
Red-necked Phalarope	*Phalaropus lobatus*	7.75 in (20 cm)
Ruff	*Philomachus pugnax*	12.5 in (32 cm)
Southern Lapwing	*Vanellus chilensis*	14.75 in (38 cm)
Spotted Redshank	*Tringa erythropus*	12.5 in (32 cm)
Tufted Puffin	*Fratercula cirrhata*	14.75 in (38 cm)
Pigeons and sandgrouse		
Banded Fruit Dove	*Ptilinopus cinctus*	13.25 in (34 cm)
Chestnut-bellied Sandgrouse	*Pterocles exustus*	13 in (33 cm)
Emerald Dove	*Chalcophaps indica*	10.5 in (27 cm)
Pallas's Sandgrouse	*Syrrhaptes paradoxus*	15.66 in (40 cm)
Pied Imperial Pigeon	*Ducula bicolor*	16 in (41 cm)
Rock Dove	*Columba livia*	13 in (33 cm)
Seychelles Blue Pigeon	*Alectroenas pulcherrima*	9.33 in (24 cm)
Victoria Crowned Pigeon	*Goura victoria*	29.66 in (76 cm)
Zebra Dove	*Geopelia striata*	8.25 in (21 cm)
Parrots		
Blue-fronted Parrot	*Amazona aestiva*	14.5 in (37 cm)
Buff-faced Pygmy Parrot	*Micropsitta pusio*	4 in (10 cm)
Burrowing Parakeet	*Cyanoliseus patagonus*	18 in (46 cm)
Eclectus Parrot	*Eclectus roratus*	14 in (36 cm)
Fischer's Lovebird	*Agapornis fischeri*	6.25 in (16 cm)
Galah	*Eolophus roseicapilla*	13.75 in (35 cm)
Ground Parrot	*Pezoporus wallicus*	11.66 in (30 cm)
Hyacinth Macaw	*Anodorhynchus hyacinthinus*	3.25 ft (1 m)
Kakapo	*Strigops habroptila*	25 in (64 cm)
Kea	*Nestor notabilis*	18.75 in (48 cm)
Maroon-faced Parakeet	*Pyrrhura leucotis*	9 in (23 cm)
Military Macaw	*Ara militaris*	27.33 in (70 cm)
Pyrrhura Leucotis	*Psittacula cyanocephala*	13 in (33 cm)
Rainbow Lorikeet	*Trichoglossus haematodus*	10 in (26 cm)
Scarlet Macaw	*Ara macao*	34.75 in (89 cm)
Senegal Parrot	*Poicephalus senegalus*	9.75 in (25 cm)
Swift Parrot	*Lathamus discolor*	9.75 in (25 cm)
White-crowned Parrot	*Pionus senilis*	9.33 in (24 cm)
Yellow-collared Lovebird	*Agapornis personatus*	5.66 in (14.5 cm)
Cuckoos and turacos		
Common Cuckoo	*Cuculus canorus*	13 in (33 cm)
Dideric Cuckoo	*Chrysococcyx caprius*	7 in (18 cm)
Great Blue Turaco	*Corythaeola cristata*	29.25 in (75 cm)
Greater Coucal	*Centropus sinensis*	20.25 in (52 cm)
Greater Roadrunner	*Geococcyx californianus*	21.75 in (56 cm)
Hartlaub's Turaco	*Tauraco hartlaubi*	16.75 in (43 cm)

Common name	Scientific name	Length/height
Jacobin Cuckoo	*Clamator jacobinus*	13.25 in (34 cm)
Rufous-vented Ground Cuckoo	*Neomorphus geoffroyi*	17.5 in (45 cm)
Smooth-billed Ani	*Crotophaga ani*	14.5 in (37 cm)
Violet Turaco	*Musophaga violacea*	19.5 in (50 cm)
Owls		
Barn Owl	*Tyto alba*	17 in (44 cm)
Barred Owl	*Strix varia*	20.66 in (53 cm)
Black-banded Owl	*Ciccaba huhula*	17.5 in (45 cm)
Boreal Owl	*Aegolius funereus*	9.75 in (25 cm)
Burrowing Owl	*Athene cunicularia*	9.33 in (24 cm)
Elf Owl	*Micrathene whitneyi*	6 in (15 cm)
Eurasian Pygmy Owl	*Glaucidium passerinum*	6.66 in (17 cm)
Great horned Owl	*Bubo virginianus*	21.5 in (55 cm)
Long-eared Owl	*Asio otus*	14.75 in (38 cm)
Northern Saw-whet Owl	*Aegolius acadicus*	7.75 in (20 cm)
Snowy Owl	*Nyctea scandiaca*	23 in (59 cm)
Spectacled Owl	*Pulsatrix perscipillata*	18 in (46 cm)
Tropical Screech Owl	*Otus choliba*	9.33 in (24 cm)
Ural Owl	*Strix uralensis*	24.25 in (62 cm)
Verreaux's Eagle-owl	*Bubo lacteus*	25.33 in (65 cm)
Nightjars and allies		
Common Pauraque	*Nyctidromus albicollis*	11 in (28 cm)
Common Poorwill	*Phalaenoptilus nuttallii*	7.75 in (20 cm)
Common Potoo	*Nyctibius griseus*	14.75 in (38 cm)
European Nightjar	*Caprimulgus europaeus*	11 in (28 cm)
Oilbird	*Steatornis caripensis*	19.5 in (50 cm)
Spotted Nightjar	*Eurostopodus argus*	11.66 in (30 cm)
Tawny Frogmouth	*Podargus strigoides*	20.66 in (53 cm)
Hummingbirds and swifts		
Alpine Swift	*Tachymarptis melba*	8.5 in (22 cm)
Asian Palm Swift	*Cypsiurus balasiensis*	5 in (13 cm)
Collared Inca	*Coeligena torquata*	5.66 in (14.5 cm)
Festive Coquette	*Lophornis chalybeus*	3.33 in (8.5 cm)
Fiery-tailed Awlbill	*Anthracothorax recurvirostris*	4 in (10 cm)
Giant Hummingbird	*Patagona gigas*	9 in (23 cm)
Gray-rumped Treeswift	*Hemiprocne longipennis*	9 in (23 cm)
Purple-throated Carib	*Eulampis jugularis*	4.66 in (12 cm)
Ruby Topaz	*Chrysolampis mosquitus*	2 in (5 cm)
Sword-billed Hummingbird	*Ensifera ensifera*	9 in (23 cm)
Mousebirds		
Speckled Mousebird	*Colius striatus*	15.66 in (40 cm)
White-headed Mousebird	*Colius leucocephalus*	13.75 in (35 cm)
Trogons		
Narina's Trogon	*Apaloderma narina*	12.5 in (32 cm)
Red-headed Trogon	*Harpactes erythrocephalus*	13.75 in (35 cm)
Resplendent Quetzal	*Pharomachrus mocinno*	15.66 in (40 cm)
White-tailed Trogon	*Trogon viridis*	11 in (28 cm)
Kingfishers and allies		
African Pygmy Kingfisher	*Ceyx pictus*	4.66 in (12 cm)
Banded Kingfisher	*Lacedo pulchella*	7.75 in (20 cm)
Belted Kingfisher	*Megaceryle alcyon*	13 in (33 cm)
Carmine Bee-eater	*Merops nubicus*	10.5 in (27 cm)

Common name	Scientific name	Length/height
Common Hoopoe	*Upupa epops*	12.5 in (32 cm)
Common Kingfisher	*Alcedo atthis*	6.25 in (16 cm)
Cuban Tody	*Todus multicolor*	4.25 in (11 cm)
Cuckoo-roller	*Leptosomus discolor*	19.5 in (50 cm)
Dollarbird	*Eurystomus orientalis*	12.5 in (32 cm)
European Roller	*Coracias garrulus*	11.66 in (30 cm)
Great Hornbill	*Buceros bicornis*	3.66 ft (1.1 m)
Green Kingfisher	*Chloroceryle americana*	8.5 in (22 cm)
Hook-billed Kingfisher	*Melidora macrorrhina*	10.5 in (27 cm)
Laughing Kookaburra	*Dacelo novaeguineae*	16.75 in (43 cm)
Lilac-cheeked Kingfisher	*Cittura cyanotis*	11 in (28 cm)
Pied Kingfisher	*Ceryle rudis*	11 in (28 cm)
Woodpeckers and allies		
Acorn Woodpecker	*Melanerpes formicivorus*	9 in (23 cm)
Black-rumped Woodpecker	*Dinopium benghalense*	11.33 in (29 cm)
Blue-throated Barbet	*Megalaima asiatica*	9 in (23 cm)
Channel-billed Toucan	*Ramphastos vitellinus*	21.75 in (56 cm)
Curl-crested Aracari	*Pteroglossus beauharnaesii*	18 in (46 cm)
Emerald Toucanet	*Aulacorhynchus prasinus*	14.5 in (37 cm)
Golden-tailed Woodpecker	*Campethera abingoni*	9 in (23 cm)
Gray Woodpecker	*Dendropicos goertae*	7.75 in (20 cm)
Gray-breasted Mountain Toucan	*Andigena hypoglauca*	18.75 in (48 cm)
Greater Honeyguide	*Indicator indicator*	7.75 in (20 cm)
Green Woodpecker	*Picus viridis*	13 in (33 cm)
Ground Woodpecker	*Geocolaptes olivaceus*	11.66 in (30 cm)
Northern Wryneck	*Jynx torquilla*	6.25 in (16 cm)
Paradise Jacamar	*Galbula dea*	13.25 in (34 cm)
Pileated Woodpecker	*Dryocopus pileatus*	18 in (46 cm)
Red-and-yellow Barbet	*Trachyphonus erythrocephalus*	9 in (23 cm)
Rufous Woodpecker	*Celeus brachyurus*	9.75 in (25 cm)
Spotted Puffbird	*Bucco tamatia*	7 in (18 cm)
Yellow-bellied Sapsucker	*Sphyrapicus varius*	8.25 in (21 cm)
Passerines		
American Goldfinch	*Carduelis tristis*	5 in (13 cm)
American Robin	*Turdus migratorius*	19.75 in (25 cm)
Asian Paradise-flycatcher	*Terpsiphone paradisi*	19.5 in (50 cm)
Australian Magpie	*Gymnorhina tibicen*	15.66 in (40 cm)
Bananaquit	*Coereba flaveola*	4 in (10 cm)
Bar-bellied Cuckoo-shrike	*Coracina striata*	12.5 in (32 cm)
Barred Antshrike	*Thamnophilus doliatus*	6.25 in (16 cm)
Black Phoebe	*Sayornis nigricans*	7.5 in (19 cm)
Black-bellied Gnateater	*Conopophaga melanogaster*	6 1.4 in (16 cm)
Black-capped Vireo	*Vireo atricapilla*	4.25 in (11 cm)
Black-crowned Sparrow-lark	*Eremopterix nigriceps*	4.25 in (11 cm)
Black-thighed Grosbeak	*Pheucticus tibialis*	7.75 in (20 cm)
Black-throated Accentor	*Prunella atrogularis*	6 in (15 cm)
Black-throated Huet-huet	*Pteroptochos tarnii*	9 in (23 cm)
Black-throated Thrush	*Turdus atrogularis*	10.5 in (27 cm)
Blue-crowned Manakin	*Lepidothrix coronata*	3.5 in (9 cm)
Blue-faced Parrotfinch	*Erythrura trichroa*	5 in (13 cm)
Bluethroat	*Luscinia svecica*	5.5 in (14 cm)
Bohemian Waxwing	*Bombycilla garrulus*	8 in (20 cm)
Brown-throated Sunbird	*Anthreptes malacensis*	5.5 in (14 cm)
Cape Sugarbird	*Promerops cafer*	18 in (46 cm)
Chestnut-crowned Babbler	*Pomatostomus ruficeps*	8.5 in (22 cm)
Collared Redstart	*Myioborus torquatus*	5 in (12.5 cm)
Crested Tit	*Parus cristatus*	4.5 in (11.5 cm)
Crimson Chat	*Epthianura tricolor*	4.66 in (12 cm)
Crimson-breasted Flowerpecker	*Prionochilus percussus*	3.5 in (9 cm)
Eastern Paradise Whydah	*Vidua paradisaea*	13 in (33 cm)
Eastern Whipbird	*Psophodes olivaceus*	10.33 in (26.5 cm)
Eurasian Golden Oriole	*Oriolus oriolus*	8.5 in (22 cm)
Fairy Gerygone	*Gerygone palpebrosa*	4.25 in (11 cm)
Fluffy-backed Tit-babbler	*Macronous ptilosus*	6.25 in (16 cm)
Goldcrest	*Regulus regulus*	3.5 in (9 cm)
Golden Bowerbird	*Prionodura newtoniana*	9.75 in (25 cm)
Golden-headed Manakin	*Pipra erythrocephala*	3.5 in (9 cm)
Gouldian Finch	*Erythrura gouldiae*	5.5 in (14 cm)
Greater Short-toed Lark	*Calandrella brachydactyla*	6 in (15 cm)
Green Broadbill	*Calyptomena viridis*	7.5 in (19 cm)
Icterine Warbler	*Hippolais icterina*	5.33 in (13.5 cm)
Madagascan Wagtail	*Motacilla flaviventris*	7.5 in (19 cm)
Marsh Wren	*Cistothorus palustris*	5 in (13 cm)
McKay's Bunting	*Plectrophenax hyperboreus*	6.66 in (17 cm)
Northern Scrub Robin	*Drymodes superciliaris*	8.5 in (22 cm)
Orange-bellied Leafbird	*Chloropsis hardwickei*	7.5 in (19 cm)
Oriental White-eye	*Zosterops palpebrosus*	4.25 in (11 cm)
Penduline Tit	*Remiz pendulinus*	4.25 in (11 cm)
Red-backed Shrike	*Lanius collurio*	7 in (18 cm)
Red-bellied Pitta	*Pitta erythrogaster*	6.25 in (16 cm)
Red-billed Blue Magpie	*Urocissa erythrorhyncha*	27.33 in (70 cm)
Red-billed Buffalo Weaver	*Bubalornis niger*	9 in (23 cm)
Red-billed Scythebill	*Campylorhamphus trochilirostris*	10.5 in (27 cm)
Red-browed Treecreeper	*Climacteris erythrops*	6.25 in (16 cm)
Red-headed Honeyeater	*Myzomela erythrocephala*	4.66 in (12 cm)
Red-throated Pipit	*Anthus cervinus*	6.25 in (16 cm)
Red-whiskered Bulbul	*Pycnonotus jocusus*	7.75 in (20 cm)
Red-winged Blackbird	*Agelaius phoeniceius*	8.5 in (22 cm)
Ruddy Treerunner	*Margarornis rubiginosus*	6.25 in (16 cm)
Satin Bowerbird	*Ptilonorhynchus violaceus*	11.66 in (30 cm)
Scarlet Minivet	*Pericrocotus flammeus*	9 in (23 cm)
Scarlet Tanager	*Piranga olivacea*	6.66 in (17 cm)
Shining Starling	*Aplonis metallica*	8.5 in (22 cm)
Southern Red Bishop	*Euplectes orix*	4.66 in (12 cm)
Spotted Pardalote	*Pardalotus punctatus*	3.5 in (9 cm)
Standardwing	*Semioptera wallacii*	10.5 in (27 cm)
Streak-chested Antpitta	*Hylopezus perspicillatus*	5.5 in (14 cm)
Superb Lyrebird	*Menura novaehollandiae*	35 in (90 cm)
Three-wattled Bellbird	*Procnias tricarunculata*	11.66 in (30 cm)
Tropical Gnatcatcher	*Polioptila plumbea*	3.5 in (9 cm)
Tufted Flycatcher	*Mitrephanes phaeocercus*	5.33 in (13.5 cm)
Turquoise Cotinga	*Cotinga ridgwayi*	7.25 in (18.5 cm)
Variegated Fairy-wren	*Malurus lamberti*	6 in (15 cm)
Wallcreeper	*Tichodroma muraria*	6.66 in (17 cm)
Western Parotia	*Parotia sefilata*	13 in (33 cm)
White-banded Swallow	*Atticora fasciata*	6 in (15 cm)
White-browed Shortwing	*Brachypteryx montana*	5 in (13 cm)
White-browed Woodswallow	*Artamus superciliosus*	7.75 in (20 cm)
White-capped Dipper	*Cinclus leucocephalus*	6.5 in (16.5 cm)
White-necked Picathartes	*Picathartes gymnocephalus*	15.66 in (40 cm)
White-necked Raven	*Corvus albicollis*	21.5 in (55 cm)
White-winged Swallow	*Tachycineta albiventer*	5.5 in (14 cm)
Yellow-bellied Fantail	*Rhipidura hypoxantha*	4.66 in (12 cm)

Glossary

Adaptive radiation Changes in a bird's anatomy, physiology or behavior to survive and breed in different geographical areas and habitats. Allows one species to evolve (or radiate) from a common ancestor into two or more different species.

Air sac Any one of a system of membranous cavities in a bird's body, forming part of the respiratory system.

Alarm call A recognizable, simple call given by an individual to warn its young or other members of the same or different species of an intruder. Elicits hiding or escape behavior.

Albumen Egg white, composed mainly of water and protein.

Altricial Recently hatched birds that are blind, featherless, and helpless at hatching.

Anti-predator behavior Any bird behavior aimed at repelling a predator or preventing a predator from finding themselves, a nest or chicks. Includes camouflage of adults, eggs or nests, mobbing and attacking predators.

Asynchronous hatching A pattern where eggs of a clutch hatch one by one over a series of days, caused by parental incubation beginning with the first-laid egg. Results in young of different sizes within a brood. Often only the oldest and largest chick survives by monopolizing the food or fratricide.

Avian Of or about birds. Birds form the class Aves in the animal kingdom.

Avian diversity The number of different species of birds within a defined geographical area.

Barb A part of the feather vane. Barbs emerge from the central shaft of a feather in parallel arrangement, like the teeth on a comb.

Barbules Small hooks that come off both sides of the barbs. Adjacent barbules hook together, keeping the vane intact.

Bill The horny covering of the jaws of a bird, comprising two halves—the maxilla (upper) and the mandible (lower). Sometimes called a beak.

Binocular vision Vision in which an object can be seen with both eyes. Birds, such as owls, that have flat faces and forward-looking eyes have a wide range of binocular vision.

Biome Large major habitat type that is generally identified with its dominant vegetation type. Biomes include grasslands, coniferous forests, deserts and tundra.

Brood The number of young birds hatched in one clutch, at one time. As a verb, to brood means to shelter young birds from the sun, heat or cold, or predators.

Brood parasite A bird that lays its eggs in the nest of another species, leaving its egg and young to be raised by the host species. A young brood parasite may eject its host's eggs or young out of the nest, thereby killing them.

Brood patch A naked, highly vascularized bare area on a bird's belly which transfers the adult's body heat to its eggs when the bird sits on them. Also called incubation patch.

Call A sound uttered by a bird that is often unconnected with either courtship or announcement of territory.

Camouflage The colors and patterns of a bird that enable it to blend with the background. Camouflage conceals birds from predators and helps them ambush prey.

Carrion The flesh of dead animals that is eaten by birds or other animals. Birds that eat carrion are called scavengers.

Class One of several divisions in which scientists divide animals. Birds are in the class called Aves.

Clutch The full set of eggs laid by a female bird in a single nesting attempt.

Colonial nester A bird that nests with others of its own species (monospecific colonies), or other species (mixed-species colonies), in close proximity.

Convergent evolution The situation in which totally unrelated birds develop similar traits to cope with similar evolutionary or environmental pressures.

Courtship The behavior patterns that male and female birds display when they are trying to attract a mate.

Crèche Gathering of young, dependent birds into a cohesive group for protection from predators, attended by one or more adults. Formed by penguins and pelicans, among others.

Cryptic Marking or coloring that makes a bird difficult for a predator to see against the bird's natural surroundings.

Dilution effect The likelihood of an individual bird being singled out by a predator is reduced in a flock or nesting colony. The larger the flock, the lower the risk for each individual bird.

Display Behavior used by a bird to communicate with its own species or with other animals. Displays can include posturing and exhibiting brightly colored parts of the body, and may signal threat, defense or readiness to mate.

Distraction display A pattern of behavior used by adults of some species to lure a predator away from its nest or chicks. Includes "broken wing" and "injury feigning" displays.

Distribution Where a species is found, including its habitat, range and location at different seasons.

Diurnal Active during the daytime.

Divergent evolution The situation in which two or more similar species become more and more dissimilar due to environmental adaptations.

DNA A molecule, found in chromosomes of a cell nucleus, that contains genes. DNA stands for deoxyribonucleic acid.

Domestication The process of taming and breeding a bird for human use. Domesticated birds include some pets, as well as birds used for sport or food.

Echolocation Navigation that relies on sound rather than sight or touch. Some birds use echolocation to tell them where they are, where their prey is and if an obstruction is in their way.

Egg The hard-shelled, reproductive unit laid by birds (both fertilized and unfertilized). An egg contains the embryo, yolk and white. The rigid but porous shell allows exchange of gases with the external environment.

Egg tooth A sharp, tooth-shaped calcium deposit that grows on the tip of the bill of an embryonic bird. The bird uses the tooth to help it break through the shell when it is hatching.

Embryo An unborn animal in the earliest stages of development. A bird embryo grows outside its mother's body, in an egg.

Endemic A species, or other taxon, found only in one habitat or region. For example, Emus are endemic to Australia.

Exotic A foreign or non-native species, often introduced deliberately or accidentally into a region by humans.

Feather The components of a bird's covering or plumage. A feather is made of keratin, and has a long shaft with two vanes on either side. The vanes, made up of many closely spaced barbs, give the feather its shape and color.

Fledgling Refers to leaving the nest, usually when able to fly. However, some birds leave the nest before they can fly.

Fossil A remnant, impression or trace of a plant or animal from a past geological age, usually found in rock.

Gizzard In birds, the equivalent of the stomach in mammals. Grit and stones inside the gizzard help to grind up food.

Gondwana Ancient southern supercontinent, comprising the present-day continents of Australia, India, Africa, South America and Antarctica.

Gregarious Showing a tendency to form flocks or to congregate with other birds.

Gullet In birds, the gullet is the equivalent of the esophagus in mammals. This tube passes food from the bill to the gizzard.

Habitat The area in which an animal naturally lives. Many different animals live in the same environment, but each kind lives in a different habitat within that environment. For example, some live in the trees; others on the ground.

Hatchling A bird that has recently hatched.

Hawking To catch and eat insects on the wing.

Herptile Any type of reptile or amphibian.

Imprinting A process in which hatchlings identify with and attach themselves to a parental figure.

Incubate To keep eggs in an environment, outside the female's body, in which they can develop and hatch. Most birds incubate their eggs by warming them with body heat.

Introduced A species imported from another place by humans and deliberately or accidentally released into a new habitat.

Iridescent colors Showing different colors as light strikes from different angles. In birds, iridescence is caused by structural properties of the feathers and their variable reflection of light.

Juvenile A young bird with its first set of functional feathers.

Keel A ridge on the breastbone that provides an attachment site for the flight muscles.

Keratin The protein from which feathers are constructed.

Kleptoparasitism Stealing food from another individual, also called piracy.

Lek An arena or place where males gather to display to females, who approach, select a male, mate, then go off and nest on their own.

Migration The regular, round-trip, movement of birds from breeding grounds to wintering grounds. Many birds migrate vast distances to find food or to breed.

Mobbing A common defense strategy employed by a group of birds encouraging predators to "move on." Birds swoop and attack predators to defend themselves, nests and young.

Molt The periodic replacement of old feathers with new ones.

Monogamous Describes male and female birds which pair to form a single couple.

Morph A color or other physical variant within a local population of a species. Giant Petrels have a rare white morph; the Snow Goose has an uncommon dark morph that is known as the Blue Goose.

Neotropics The New World tropical zones; includes all of South America, and Central America to southern Mexico.

New World North and South America.

Niche The ecological position occupied by a species within an animal community.

Nomad A bird which lacks fixed territory, and wanders instead from place to place in search of food and water.

Old World Eurasia, Africa and Australasia.

Omnivore A bird that eats both plant and animal food.

Order A major group used in taxonomic classification. An order forms part of a class, and comprises one or more families.

Oscine A passerine that has a complex syrinx which allows it to sing complex songs.

Pair bond A partnership maintained between a male and a female bird through one or several breeding attempts. Some species maintain a pair bond for life.

Passerine Any species of bird belonging to the order Passeriformes. Also known as perching birds.

Peripheral vision Ability to see to the sides, even slightly backward in some birds.

Philopatry Refers to returning to the same area to breed, such as a colony site. Also refers to returning to the same exact nest site in successive years, which occurs in most colonial birds and some hawks, owls and parrots.

Phylogeny Relationships among different groups of birds, based on evolutionary descent.

Pigment Any substance that creates color in the skin, feathers or tissues of a bird, other animal or plant.

Plumage The sum total of feathers on a bird's body.

Polyandry A female having more than one male mate.

Polygamous Describes birds that have more than one mate.

Polygyny A male having more than one female mate.

Precocial Active and self-reliant at birth. Describes newly hatched chicks of some birds, such as ducks and chickens.

Predator A bird or other animal that lives mainly by killing and eating other animals.

Preen To clean, repair, arrange and maintain plumage.

Rain forest A tropical forest that receives at least 100 inches (250 cm) of rain each year.

Range The entire geographic area across which a species is regularly found.

Raptor A diurnal bird of prey, such as a hawk or falcon. The term is not usually used to describe owls.

Regurgitate To bring food back up from the crop to feed to young or to remove foreign objects from the digestive tract.

Rhamphotheca Horny outer covering (sheath) of a bird's bill.

Roost A place or site used by birds for sleeping. Also, the act of settling at such a place.

Sallying To forage by flying out from a perch to pick up food from surfaces or in the air, then returning to a perch to eat.

Savanna Open grassland with scattered trees. Most common in subtropical areas that have a distinct summer wet season.

Scavenger A bird that eats carrion—often the remains of animals killed by predators.

Sedentary Having a lifestyle that involves little movement; also used to describe birds that do not migrate.

Sexual dimorphism The case in which the male and female of a species differ in appearance or size.

Shaft The long, slender, central part of the feather that holds the vanes (also called a rachis).

Social Living in groups. Social birds can live in breeding pairs, sometimes together with their young, or in colonies of up to thousands of birds.

Song Any vocalization of a bird with the particular purpose of obtaining a mate, announcing a territory, or warning intruders. Not all birds sing.

Speciation The evolutionary process in which populations which are isolated evolve into separate species, and are no longer able to interbreed.

Species A population of birds with very similar features that are able to breed together and produce fertile young.

Sternum The breastbone. Flying birds have a large, deeply keeled sternum to anchor their powerful flight muscles.

Stop-over A location along a migration route where birds stop to refuel (regain weight), to allow further migration.

Subantarctic Of the oceans and islands just north of Antarctica.

Suboscine A suborder within the passerines that has a less complex syrinx than oscines. Not considered true songbirds.

Syrinx The organ that produces sound, located where the trachea divides to form two bronchi. Changes in the shape and diameter of the syrinx alters the sound emitted, producing the calls and sounds of birds.

Taxonomy The science of classifying living things into groups and subgroups according to similarities and adaptations.

Temperate An environment or region that has a warm summer and a cool winter. Most of the world's temperate regions are located between the tropics and the polar regions.

Territory An area defended by a bird for its own exclusive use against intruders of its own species (or, occasionally others). The area often incorporates all living resources needed by the bird, such as food and a nesting and roosting site.

Thermal A column of rising air used by birds to gain height. Some birds soar on thermals to save energy before gliding downward again.

Torpid In a sleeplike state in which bodily processes are slowed. Torpor helps birds to survive difficult conditions such as cold or lack of food. Estivation and hibernation are types of torpor.

Trachea The windpipe or tube allowing air from the mouth to reach the lungs.

Transient A migrant bird in transit across the area between its normal breeding and wintering distributions.

Tropical Environments or regions near the Equator that are warm to hot all year round.

Tundra A cold, barren habitat found near the Arctic Circle and on mountain tops, at high altitude or high latitude.

Vagrant Any bird outside the normal distribution of its species.

Vane The plumed part of the feather that grows from the central shaft.

Vestigial Relating to a body part that is non-functional or atrophied, such as vestigial wings in ratites.

Vortex Circular air currents made by wingtip movement.

Wing loading The ratio between a bird's weight and the combined surface area of the wing.

Zygodactyl feet Where two toes point forward and two point backward. Improves manipulation and grasping of objects. Present in birds such as woodpeckers, cuckoos and most parrots.

Index

D

E

F

G

H

I

J

S

T

Credits and acknowledgments

PHOTOGRAPHS

Key t=top; l=left; r=right; tl=top left; tcl=top center left; tc=top center; tcr=top center right; tr=top right; cl=center left; c=center; cr=center right; b=bottom; bl=bottom left; bcl=bottom center left; bc=bottom center; bcr=bottom center right; br=bottom right

APL = Australian Picture Library; APL/CBT = Australian Picture Library/Corbis; APL/MP = Australian Picture Library/Minden Pictures; AUS = Auscape International; GI = Getty Images; MEPL = Mary Evans Picture Library; NHPA = Natural History Photographic Agency; NPL=Nature Picture Library; PD = Photodisc; PL = photolibrary.com

Front Cover t NHPA; bl APL/CBT; bcl GI; bcr APL/CBT; br GI **Front Flap** AUS **Back Cover** APL/MP **Spine** GI **1**c GI **2**c GI **4**r PL **6**c GI **8**c APL/CBT **10**c APL/CBT l, r APL/MP **11**c AUS l GI r APL/CBT **12**c GI **14**c APL/MP **16**cl, r APL/MP cr PL l APL/CBT **17**c APL/CBT **18**br PL t NHPA **19**bl, br PL **20**tr APL/MP **21**cl APL/MP r, tr APL/CBT **22**b, c APL/MP **24**bl PL br NHPA r APL/MP **25**br AUS **26**bl PL **29**br PL **30**r APL/MP **31**bl NHPA **32**b NHPA tr PL **33**bl NHPA tr APL/MP **34**bl NHPA br, tr APL/MP **35**br GI t APL/MP **36**b APL/MP **37**bl PL br, tr APL/MP cl, tl NHPA **38**bl, tr PL **39**bl APL/MP br, tr GI **40**b PL **41**bl, tr APL/MP bl NHPA **42**bl NHPA tr APL/CBT **43**bl APL/CBT tl PL tr APL/MP **44**l APL/MP **45**b APL/MP tc GI tl, tr PL **46**b APL/MP **47**tl, tr APL/MP **48**bl APL/CBT r GI **50**bl, r APL/MP **51**tl APL/MP **52**b APL/MP **53**bl NHPA br APL/MP t GI **54**b, t APL/MP **55**b APL/MP **56**c GI c APL/CBT **58**cl PL cr, l APL/MP **59**l, r GI **60**bl, br APL/MP tr APL/CBT **61**bl APL/MP r PL **62**b APL/MP tr GI **63**br PL cr, tr APL/MP **64**b PL tr GI **65**b APL/MP t GI **66**bl PL br APL/MP **67**r, tl APL/MP **68**t APL/MP **69**b, t APL/MP **70**bc, br NHPA tr GI **71**bc, bl, br NHPA t GI **72**b PL tr APL/MP **73**br, tl GI **74**l APL/CBT **75**b, tl APL/MP tr PL **76**tr APL/MP **77**r, tl APL/MP **78**b NHPA tr GI **79**bl GI tr NHPA **80**bl, tr APL/MP **81**bl APL/MP **82**b APL/MP tr Borror Laboratory of Bioacoustics, Department of Evolution, Ecology, and Organismal Biology, Ohio State University, Columbus, OH, all rights reserved. **83**r GI **84**bl APL/MP cr NPL tr PL **85**c PL **86**bl NHPA tr APL/MP **87**b APL/MP tr NHPA **88**cr NHPA tr PL **89**r GI t APL/MP **90**b PL **91**br PL tl APL/MP tr GI **92**bl, tr APL/MP cr GI **93**c GI **94**b NHPA tr APL/MP **95**bl PL br APL/MP tl GI **96**bl NHPA r APL/MP **97**l GI tr NHPA **98**r APL/MP **99**br APL/MP tl NHPA tr PL **100**br PL **101**c NHPA **102**r AUS **104**c APL/MP **106**c NHPA l APL/CBT r GI **107**c APL/MP l NHPA **108**cl APL/MP tr APL/CBT **109**cl PL **110**r PL **111**b, tl APL/MP **112**b APL/MP t PL **113**cr NHPA l APL/CBT tr APL/MP **114**b APL/MP **115**b, c GI tl PL **116**b NHPA **117**br, tl APL/MP **118**bl APL/MP tr PL **119**br, t GI **120**b PL tr NHPA **121**cr, tl APL/MP **122**b, tr APL/MP **123**b APL/CBT **124**br APL/MP t PL **125**br PL tl, tr GI **126**b GI **127**br, tl, tr APL/MP **128**r APL/CBT **129**bl APL/CBT cr APL/MP tr NHPA **130**b, tr GI **131**c GI **132**b GI **133**bl, br, tr GI **134**b PL **135**b APL/MP **136**br GI tl APL/MP tr APL/CBT **137**br, c NHPA tr GI **138**r APL/MP **139**tl GI tr APL/MP **140**c PL **141**bc NHPA c APL/CBT r APL/MP **142**bc, tr PL cr AUS **143**b NHPA **144**bl APL/CBT c PL **145**br PL cr NHPA l APL/MP **146**bc APL/MP tr AUS **147**bc PL br APL/CBT tr APL/MP **148**r APL/MP **149**br APL/CBT l NHPA tr GI **150**br APL/MP tr APL/CBT **151**b APL/MP **152**t APL/MP **153**br, tl APL/CBT tr PL **154**r APL/MP **155**br APL/CBT tl PL tr APL/MP **156**bc GI br APL/CBT **157**c PL **158**b APL/MP **159**br APL/MP br, tl APL tr PL **160**bl APL/MP r GI **161**br GI cr PL l APL/MP **162**b PL tl APL/CBT **163**br NHPA tr PL **164**bl, tl APL/MP br NHPA **165**t APL/MP **166**b PL **167**b APL/CBT t APL/MP **168**bl APL/MP r APL/CBT **169**b, tl PL **170**b, tr APL/MP **171**c APL/MP **172**c GI **174**c APL/MP l GI r NHPA **175**c PL l APL/MP **176**tr APL/CBT **177**bl, tr PL br APL **178**br, tr APL/MP **179**bl NHPA tr GI **180**b, tr APL/MP **181**br GI **182**br, tr GI **183**br APL/MP tr NHPA **184**b GI tr PL **185**r APL/MP **186**r APL/MP **187**r APL/MP tl GI **188**b APL tr APL/MP **189**tr APL/MP **190**bl AUS r NHPA **191**br APL/MP tr GI 192bl PL br APL/MP tr NHPA **193**r NHPA **194**r APL/MP **195**br PL tr NHPA **196**bl, br PL **197**b NHPA tl APL/CBT **198**bl PL r APL/MP **199**cr, tr APL/MP **200**br NHPA tr APL/MP **201**b NPL tr NHPA **202**b PL tr APL/MP **203**r GI **204**br NHPA cl APL/MP tr NPL **205**r NHPA **206**bl, r APL/MP **207**tl GI **208**r APL/CBT tl PL **209**tc APL/MP tr Tom Stephenson **210**c AUS **212**c GI l, r APL/MP **213**c PL l NHPA **214**b APL/MP **215**br NHPA l GI **216**br NHPA tr PL **217**br APL/MP **218**b APL/MP **219**bl APL/MP tl NHPA tr AUS **220**bl APL/MP br NHPA **221**b PL t GI **222**b APL/MP **223**bl NHPA tl PL **224**cl NHPA cr PL **225**bl, tl GI **226**bl PL r NHPA **227**br APL/CBT tl GI **228**b APL/MP **229**bl, br APL/MP tl AUS **230**b APL/MP **231**bl APL/MP br, tl NHPA **232**bl GI tr NHPA **233**bl NHPA **235**c APL/MP **236**bl NHPA **237**t APL/MP **238**bl NHPA **239**bl NHPA br APL/CBT t PL **240**br NHPA t APL/MP **241**b, t PL **242**b, tr APL/MP **243**br PL tl, tr APL/MP **244**c APL/CBT **246**c APL/MP l, r GI **247**c, cr GI l APL/MP **248**b NHPA tl GI **249**b APL/MP t GI **250**bl, tr GI br APL/MP **251**br AUS **252**b APL/MP **253**bl National Library of Australia tr PL **254**bl, br PL tr APL/MP **255**c GI **256**bl GI br APL/CBT **257**br APL/MP t NHPA **258**bl, r APL/MP **259**cr NHPA tl, tr GI **260**b APL/MP **261**cr APL/MP tl NHPA tr GI **262**b GI **263**l GI tr PD **264**br NHPA tr APL/MP **265**c GI **266**bl, r GI **267**tr GI **268**bl NHPA r APL/MP **269**br Joanna Burger tl, tr APL/MP **270**bl NHPA cr APL/MP tr GI **271**br NHPA t APL/MP **272**b GI **273**cr, l APL/MP tr GI **274**b APL/MP **275**br PL t APL/MP **276**bl GI br MEPL tr PL **277**r GI **278**bl, tr PL br GI **279**b APL/CBT tr GI 280r GI **281**bl GI br, tl PL tr MEPL **282**b Robert Bateman tr APL/CBT **283**r APL/CBT **284**c APL/CBT

ILLUSTRATIONS

Andrew Davies/Creative Communication and Map Illustrations 110bl, 112bl, 114tr, 116tr, 118cl, 120bl, 121tr, 122bl, 123tr, 126tr, 128bl, 132tr, 134tr, 137bl, 138bl, 142bl, 143tr, 146bl, 147bl, 150bl cl, 151tr, 154bl, 156tr, 158tr*, 176b, 178bl, 179br, 181tl tr

Andrew Davies/Creative Communication108bl, 109c, 259bl

The Art Agency 20b, 27r, 28cl, 30bl, 31tl, 51b, 55t, 58r, 68bl, 70c, 76b, 83b, 117bl, 130l, 156cl, 207r, 224bl br, 225tr, br, 234r

Gino Hasler 26cr, 28br

David Kirshner 115tr, 218tr, 253br tc, 286l, 289br, 290l

MagicGroup s.r.o. (Czech Republic) - www.magicgroup.cz 122tr, 133tr, 148bl,

Rob Mancini 49r, 129cl, 223r

Map Illustrations 103br cr tr

ACKNOWLEDGMENTS

The publisher wishes to thank Glenda Browne for the index and Michael Gochfeld for extensive technical research contributed to this project.

* Map originally from Ericson, P.G.P., Christidis, L., Cooper, A., Irestedt, M., Jackson, J., Johansson, U.S. & Norman, J.A. *A Gondwanan origin of passerine birds supported by DNA sequences of the endemic New Zealand wrens.* Proceedings of the Royal Society of London. Ser. B., 269, pp. 235–241; modified by Andrew Davies/Creative Communication

CAPTIONS

page 1 To bathe, a Greylag Goose dips its head underwater, then shakes it to splash water over its body.

page 2 Hundreds of Green Parrots inhabit Golondrinas, a deep vertical cave in San Luis Potassi State, Mexico. At daybreak, they exit en masse to search for food.

pages 4–5 Long-eared Owls inhabit forests where they perch in trees. Their cryptic, mottled brown plumage allows them to blend in with tree trunks.

pages 6–7 Flamingos forage in Kenya. Their pink coloring is due to their carotinoid-rich diet, which is largely made up of algae and assorted types of insects.

pages 8–9 During summer, the Willow Ptarmigan's brown feathers provide camouflage on the grassy tundra. In winter, its feathers turn white to blend in with its snowy surrounds.

pages 12–13 Birds inhabit all parts of the globe, including some of the most inhospitable areas. Here, a Southern Giant Petrel flies into a storm in Antarctica.

pages 284–5 Birds' eggs are as varied as the different kinds of birds that lay them. Some are camouflaged for maximum protection, others are brightly colored and patterned.